To Dan

Night To Day

Alabama Recipients of the Congressional Medal of Honor

John L. Johnson, Ph.D.

Tristan Press
Demopolis, Alabama

TO ALL MY LOVES:
Susan, Tristan, and Julian, and special thanks to Elie

Published by:

TRISTAN PRESS
P. O. Box 901
Demopolis, Alabama 36732 (USA)

NIGHT TO DAY

Interior photos courtesy of www.homeofheroes.com
Book design by Rashad Milhouse of Selma, Alabama

This book is an original publication of TRISAN PRESS.

First Edition: March 2022

ISBN: 979-8-9857124-0-7

An application to register this book for cataloging has been submitted to the Library of Congress.

PRINTED IN THE UNITED STATES OF AMERICA

10 9 8 7 6 5 4 3 2 1

Every Night and Every Morn
Some to Misery Are Born
Every Night and Every Morn
Some are Born to Sweet Delight
Some are Born to Sweet Delight
Some are Born to Endless Night
God Appears and God is Light

(From "Auguries of Innocence" by William Blake)

TABLE OF CONTENTS

ACKNOWLEDGMENTS

I wish to acknowledge and thank a number of individuals and organizations that made this book possible. First, I wish to thank the National Council of La Raza and the Cherokee Nation for their financial support. When solicited for financial help to begin this book, these two organizations quickly and generously stepped forward.

I also wish to thank the Congressional Medal of Honor Society and especially Victoria Kueck, Carol Cepregi, Laurie Jowdy, and Timothy Frank for their advice and help and especially for making available the Society's invaluable archives.

I gratefully acknowledge those who read and provided advice and valuable technical assistance in the development of the book, including Dr. Julie Edelson, for her excellent editorial guidance and assistance, Rashad Milhouse of Selma, Alabama for design and layout, Mr. Rudy Anderson for media relations, and Scott Crockett of Keiger Printing. All of these friends and colleagues gave generously of their time and knowledge. I wish to thanks to Wikipedia and *homeofheroes.com* for giving permission to use some of their photographs.

37TH CONGRESS,
2D SESSION.

S. R. 82.

IN THE SENATE OF THE UNITED STATES.

MAY 13, 1862.

Mr. WILSON, of Massachusetts, from the Committee on Military Affairs and the Militia, reported the following joint resolution; which was read and passed to a second reading.

JOINT RESOLUTION

To provide for the presentation of "medals of honor" to the enlisted men of the army and volunteer forces who have or who may distinguish themselves in battle during the present rebellion.

Resolved by the Senate and House of Representatives of the United States of America in Congress assembled, That the President of the United State be, and he is hereby, authorized to cause two thousand "medals of honor" to be prepared with suitable emblematic devices, and to direct that the same be presented, in the name of Congress, to such non-commissioned officers and privates as shall most distinguish themselves by their gallantry in action, and other soldier-like qualities, during the present insurrection. And that the sum of ten thousand dollars be, and the same is hereby, appropriated out of any money in the treasury not otherwise appropriated, for the purpose of carrying this resolution into effect.

ALABAMA MEDAL OF HONOR FACTS

TOTAL ALABAMA RECIPIENTS	33
TOTAL: ARMY	22
TOTAL: NAVY	5
TOTAL AIR FORCE	1
TOTAL: MARINES	5
TOTAL: COAST GUARD	0
FIRST ALABAMA RECIPIENT	MARCH 25, 1863 TO CORPORAL WILLAM REDDICK, COMPANY D, 33RD OHIO INFANTRY
NO. ALABAMA AFRICAN AMERICAN RECIPIENTS	3
NO. CURRENT LIVING ALABAMA RECIPIENTS	1

ALABAMA MEDAL OF HONOR RECIPIENTS BY CONFLICTS

AMERICAN CIVIL WAR	3
SPANISH-AMERICAN WAR	1
INTERIM PERIOD	1
WORLD WAR I	2
WORLD WAR TWO	12
KOREAN WAR	3
VIETNAM WAR	6

CHAPTER 1

INTRODUCTION

Why A Book About Alabama's Medal of Honor Recipients?

One day, while standing on a corner in downtown Birmingham waiting for the street light to change, I overhead several high school students having a conversation about career opportunities in the military. It was soon apparent they had little knowledge about the military, as their discussion was full of misconceptions and outright errors. The most prominent emerged when one of the students expressed serious doubts about whether any person from Alabama had ever received the Congressional Medal of Honor, the country's highest award for military valor. The student's misstatement was quickly accepted as fact by his friends as they boarded the bus. I wanted to intervene and tell them they were wrong and that many Alabamians had served proudly in all branches of the armed services and in all of the country's wars and that many were awarded the Medal of Honor, but it was too late—the bus pulled away.

Watching it leave, I wondered how many other young men and women were ignorant of the fact that more than 33 individuals with close connections to and affiliation with the state of Alabama had received the Medal of Honor dating back to the Civil War.

In the days that followed, I asked every high school and college student I met what they knew about the Medal of Honor and its recipients. I was disappointed to learn most knew little or nothing at all about the award. A large number had never heard of the Medal of Honor or knew what it represented. That was the day I received my *calling* to prepare this book.

This book attempts to identify and to profile all known individuals with direct or indirect affiliations to Alabama that were awarded the Medal of Honor since 1863, when it was first established by Congress during the American Civil War. My hope is that the book fulfills three main goals. First, it should serve as a reliable educational tool and resource guide that delineates the roles, deeds, and sacrifices Americans of various ethnic and religious minority groups have made in service to their country during time of war. Second, by learning about the actions that resulted in the award of the Medal, I hope that we come to honor their bravery and valor individually for the sacrifices made in the defense of their country. Last, I want to put a human face on these individuals, many of whom are like us—simple, ordinary people who found themselves in extraordinary and unimaginable circumstances, which, in turn, resulted in extraordinary and unimaginably heroic actions.

The main body of this book presents portraits of each Medal recipient. Each profile includes a photograph, if available; surprisingly, photos were available for most, including recipients from the Civil War period. As anticipated, photos of earlier periods are usually of poorer quality. Each profile also contains basic biographical and military service information: generally, the recipient's military **Rank and organization** or unit, **branch of service, date and place of the action** where the Medal was earned, and other **biographical data**. Each profile also contains the *Citation*, which describes in detail the actions and circumstances for which the Medal was awarded.[5]

The last component of the profile, the ***Epilogue***, summarizes what is known about recipients and their lives prior to, during, and after receiving the Medal. Some of these narratives are extensive, while for others, we know nothing or next to nothing. For example, as expected, information was very limited or nonexistent for recipients from the Civil War period. During the war, the military had only the most essential information for these men, including enlistment and medical records. After the war, recipients melted back into a highly segregated society, and most became "lost to history." I elaborate on the problems of lack of information in the section below, called *Limitation*.

Method

In preparing this book, I had very few resources, which presented several major obstacles. I could not travel to military and government archives, military bases, federal and state court houses, libraries, cemeteries, schools, museums, or presidential libraries to review documents and records. Nor could I interview recipients, relatives, friends, and associates to collect information.

Instead, I relied on several source: (1) the archives and files of the Congressional Medal of Honor Society (CMOHS). The CMOHS is located on a historic decommissioned navy aircraft carrier, the USS Yorktown, moored in the harbor of Charleston, South Carolina. A nonprofit organization that receives no government financial support, one of its primary missions is to serve as a repository of information on each Medal of Honor recipient.

These archives proved an invaluable resource, although the amount of information for each recipient varies widely. Most files contained a wealth of information, including copies of enlistment records Medal of Honor recommendation forms, enlistment records, court documents, bonus applications, medical records, notes on scraps of paper, newspaper clippings, letters from recipients and/or their families, friends, and associates, and research notes prepared by historians, researchers, and authors. Other recipient files, however, had minimal information and, in most cases, just a single form (Recipient Data Form), indicating that the individual had received the Medal.

I also relied heavily on several online sources and databases including Wikipedia and the Department of defense which held a wealth of information on Medal of Honor recipients.

Not only was the quantity of information vary considerably, the physical quality was also inconsistent, usually related to the quality of the photocopier used, illegible handwriting, and missing information, such as magazine articles and newspaper clippings.

The contents of each recipient's file were photocopied and analyzed, and the result of the analysis used to prepare this book.

Summary and Findings

Reviewing each recipient's file was like taking a journey back in time. Documents reveal that most Medal recipients were ordinary Americans from small towns and traditional families, who entered the military to serve their country and to seek better opportunities for themselves and their families.

The most significant and moving aspect of reviewing the files was to read the first-hand accounts of extraordinary and selfless acts of heroism that would eventually lead these men and one female to receive the Medal of Honor.

Limitations

This book has several limitations. First, the information used to develop the portraits is derived from primarily one source, the archives of the Congressional Medal of Honor Society. It is always preferable to have as many primary sources as possible when conducting research. However, lack of resources prevented conducting interviews or access to other data sources, such as the National Archives; Department of Defense Archives; National Military Personnel Center; cemetery, court, school, and church records.

Second, while some recipient files contain a wealth of information, most do not. Consequently, the depth and breadth of the narratives for some recipients cannot meet my original goal of presenting full descriptions for all.

Third, the numerous errors found in the source documents and their poor quality could not be changed, as they constitute the *official record*, which, by definition, cannot be corrected or updated. The poor quality of some documents, caused by age, defective photocopy machines, and/or illegible handwriting made information hard to discern and sometimes impossible to report. Last, there were many instances in recipient files when newspaper clippings, magazine articles, and other publications lacked proper citations; for example, copies of newspaper articles were missing authors, dates, and sources.

I hope to address these and other limitations of the book in future by enlisting the help of you, the reader.

A Lifelong Project & Legacy

My chief goal when I began this book was to prepare highly detailed, descriptive, and informative narratives for each Medal of Honor recipient. Unfortunately, this goal could not be realized for the reasons noted above and the lack of financial resources to collect information from sources that are widely dispersed throughout the world. To address this problem, I hope to publish future editions with updated portraits for each Medal recipient with information received from *you*!

I want to enlist each and every one of you who read this book to contribute to future editions. Conduct research on the life of a Medal recipient and submit your findings to me for inclusion in the next edition. You may choose to investigate a recipient who hails from your hometown or state or attended the same school. You may write about a recipient who served in the same military unit as you did. You may also consider making your submission a group project by asking friends, your class, or organization to collaborate with you.

What should your submission or narrative contain? Any personal information about the recipient will be helpful: details about birth, parents' names and occupations; schools (elementary through college); employment (positions/titles, dates); military service; spouses' names and occupation; children (gender, ages); activities (sports, volunteering, hobbies); and memorable quotes. Possible sources of information might include newspaper articles, report cards, interviews of recipients and their families, friends and associates. No information is unimportant.

You do not have to be a professional writer. You can submit short notes, outlines, or full articles. Particularly important would be photos of recipients and their families. When you are ready, forward your submission to:

John Johnson
P. O. Box 901
Demopolis, Alabama 36732
Email: jlj@tristanpress.net

I hope to publish the next edition by March 5, 2030.

CHAPTER 2

HISTORY OF THE MEDAL OF HONOR

(Excerpts from the US Department of Defense)

"The Medal of Honor is the highest military decoration awarded by the United States. It is often referred to as the *Congressional Medal of Honor* because the President awards it in the name of the Congress. It is bestowed on members of the United States armed forces who conspicuously distinguish themselves by gallantry and intrepidity at the risk of their lives, above and beyond the call of duty, while engaged in action against an enemy of the United States.

All branches of the US military are eligible to receive the Medal, and each service has a unique design, with the exception of the Marine Corps and Coast Guard, which both use the Navy's medal. The Medal of Honor is often presented personally to the recipient or, in the case of posthumous awards, to survivors, by the President of the United States. Due to its high status, the medal has special protection under US law.

The first formal system for rewarding acts of individual gallantry by American soldiers was established by George Washington on August 7, 1782, when he created the Badge of Military Merit, designed to recognize "any singularly meritorious action." This decoration is America's first combat award and the second oldest American military decoration of any type, after the Fidelity Medallion, which was awarded only once and is viewed as commemorative.

Although the Badge of Military Merit fell into disuse after the Revolutionary War, the concept of a military award for bravery had been established. In 1847, after the outbreak of the Mexican-American War, a Certificate of Merit was established for soldiers who distinguished themselves in action. The certificate was later granted medal status as the Certificate of Merit Medal.

Early in the Civil War, a medal for individual valor was proposed to Winfield Scott, the commanding general of the United States Army. He did not approve the proposal, but the medal did come into use by the Navy. Public Resolution 82, containing a provision for a Navy

Medal of Valor, was signed into law by President Abraham Lincoln on December 21, 1861. The medal was to be bestowed upon petty officers, seamen, landsmen, and Marines who "most distinguished themselves by their gallantry and other seamanlike qualities during the present war." Secretary of the Navy Gideon Welles directed the Philadelphia Mint to design the new decoration. Shortly afterward, a resolution of similar wording was introduced on behalf of the Army and signed into law on July 12, 1862. This measure provided for awarding a Medal of Honor, as the Navy version also came to be called: "to such noncommissioned officers and privates as shall most distinguish themselves by their gallantry in action, and other soldier-like qualities, during the present insurrection."

Appearance

The appearance of the Medal of Honor has evolved since its creation in 1862. The present Army medal consists of a gold star surrounded by a wreath, topped by an eagle on a bar inscribed with the word "Valor." The medal is attached by a hook to a light blue, moiré silk neckband that is 1 3⁄16 inches wide and 21¾ inches long.

There is a version of the medal for each branch of the US armed services: the Army, Navy, and Air Force. Since the US Marine Corps is administratively a part of the Navy, Marines receive the Navy medal. Before 1965, when the US Air Force design was adopted, members of the US Army Air Corps, US Army Air Forces, and Air Force received the Army version of the medal.

The Coast Guard Medal of Honor, which was distinguished from the Navy medal in 1963, has never been awarded, partly because the US Coast Guard is subsumed into the US Navy when war is declared. No design yet exists for it. Only one member of the Coast Guard has ever received a Medal of Honor: Signalman 1st Class Douglas Munro, who was awarded the Navy version of the medal for action during the Battle of Guadalcanal.

In the 19 cases where a service member has been awarded more than one Medal of Honor, current regulations specify that an appropriate device be centered on the Medal and ribbon. The Army and Air Force bestow an oak leaf cluster, while the Navy Medal of Honor is worn with a gold star.

A ribbon that is the same shade of light blue as the neckband and includes five white stars, pointed upward, in the shape of an "M" is worn for situations other than full dress uniform. When the ribbon is worn, it is placed alone, ¼ inch above the center of the other ribbons. For wear with civilian clothing, a rosette is issued, instead of a miniature lapel pin that usually shows the ribbon bar. The rosette is the same shade of blue as the neck ribbon and includes white stars. The ribbon and rosette are presented at the same time as the Medal.

The Medal of Honor can be awarded in one of two ways. The first and most common is nomination by a service member in the chain of command, followed by approval at each level of command. The other is nomination by a member of Congress, generally at the request of a constituent, and approval by a special act of Congress. In either case, the Medal of Honor is presented by the President on behalf of the Congress. Although commonplace, the term "Congressional Medal of Honor" is not correct. The Congressional Medal of Honor Society was named by an act of Congress, signed into law by President Eisenhower on August 5, 1958, as 36 U.S.C. Section 33.

Evolution of Criteria

A year after President Abraham Lincoln signed Public Resolution 82 into law on December 21, 1861; a similar resolution for the Army was passed. Six Union soldiers who hijacked a Confederate locomotive were the first recipients. Raid leader James J. Andrews, a civilian hanged as a Union spy, did not receive the medal because it was originally awarded only to enlisted men. Army officers first received them in 1891 and Navy officers in 1915. Many Medals of Honor awarded in the nineteenth century were associated with saving the flag, not just for patriotic reasons, but because it was a primary means of battlefield communication.

During the Civil War, no other military award was authorized, which explains some of the less notable actions that were recognized by the Medal of Honor. Secretary of War Edwin Stanton promised a Medal of Honor to every man in the 27th Maine Regiment who extended his enlistment beyond the agreed upon date. Many stayed four days extra and were then discharged. Due to confusion, Stanton awarded a Medal of Honor to all 864 men in the regiment.

The criteria for award tightened after World War I. In 1916, a board of five Army generals convened by law to review every Army Medal of Honor awarded. The commission, led by Nelson Miles, recommended that the Army rescind 911 medals, including the 864 to members of the 27th Maine; 29 to Abraham Lincoln's funeral guard; six civilians, including Dr. Mary Edwards Walker, the only female to have been awarded the Medal, and Buffalo Bill Cody, and 12 others whose awards were judged frivolous. Dr. Walker's medal was restored posthumously by President Jimmy Carter in 1977. Cody's award was restored in 1989.

Early in the twentieth century, the Navy awarded many Medals of Honor for peacetime bravery. For instance, seven sailors aboard the USS Iowa received the Medal when a boiler exploded on January 25, 1904. Aboard the USS Chicago in 1901, John Henry Helms received the medal for saving Ishi Tomizi, the ship's cook, from drowning. Even after World War I, Richard Byrd and Floyd Bennett received the medal for exploration of the North Pole. Thomas John Ryan received it for saving a female from the burning Grand Hotel in Yokohama, Japan, following the 1923 Great Kantō earthquake.

Since the beginning of World War II, the medal has been awarded for extreme bravery beyond the call of duty while engaged in action against an enemy. Arising from these criteria, approximately 60 percent of the medals earned during and after World War II have been awarded posthumously. Captain William McGonagle is an exception to the enemy-action rule, earning his medal during the USS *Liberty* incident, which the Israeli government claimed was fire. In the post-World War II era, many eligible recipients might instead have been awarded a Silver Star, Navy Cross, or similar award.

Controversies

A 1993 study commissioned by the Army described systematic racial discrimination in the criteria for awarding medals during World War II. At the time, no Medals of Honor had been awarded to black soldiers. After an exhaustive review of files, the study recommended that several black Distinguished Service Cross recipients be upgraded to the Medal of Honor. On January 13, 1997, President Bill Clinton awarded the medal to seven African American World War II veterans, including two from Alabama: **Private George Watson of**

Birmingham and **Lieutenant Charles L. Thomas, also of Birmingham**. Both awards were posthumous.

Privileges

The Medal of Honor confers special privileges on its recipients, both by tradition and law. By tradition, all other soldiers, sailors, Marines, and airmen—even higher ranking officers up to the President of the United States—initiate the salute. In the event of an officer encountering an enlisted member of the military who has been awarded the Medal of Honor, officers salute, not the person, but the medal itself, attempting to time their salute to coincide with the enlisted members'. By law, recipients have several benefits:

- All Medal recipients are entitled to be entered on the Medal of Honor Roll. As a result, each Recipient is entitled to receive a special pension of $1,027 per month.
- Enlisted recipients are entitled to receive a supplemental uniform allowance.
- Recipients are entitled to travel on military aircraft when space is available.
- Recipients are entitled access to commissary and exchange privileges.
- Children of recipients are eligible for admission to service academies without regard to quotas.
- Recipients after 2002 are entitled to receive the Medal of Honor Flag.
- Recipients are entitled to wear the Medal with appropriate civilian clothes.

Legal Protection

Until late 2006, the Medal of Honor was the only service decoration singled-out for federal legislative protection under the Stolen Valor Act of 2005. Now, any false verbal, written, or physical claim to an award or decoration authorized for wear by military members or veterans is a federal felony.

All Medals of Honor are issued in the original only by the Department of Defense to a recipient. Misuse of the medal, including unauthorized manufacture or wear, is punishable by fine and imprisonment and a harsher penalty is prescribed than for violations concerning other medals. After the Army redesigned its medal in 1903, a patent was issued to legally prevent others from making it. When the patent expired, the federal government enacted a law making it illegal to produce, wear, or distribute the Medal of Honor without proper authority. Violators of this law have been prosecuted. In 2003, Edward Fedora and Gisela Fedora were charged with Unlawful Sale of a Medal of Honor. They sold medals awarded to US Navy Seaman Robert Blume for action during the Spanish-American War and US Army First Sergeant George Washington Roosevelt for action during the Civil War to an FBI agent. Edward Fedora, a Canadian businessman, pleaded guilty and was sentenced to prison; Gisela Fedora's status is unknown.

While it is illegal to wear the Medal of Honor without authorization, someone can still legally claim to be a recipient unless the claim is made with the intent of securing veteran benefits. A number of veterans' organizations and private companies devote themselves to exposing those who falsely claim to have received the Medal of Honor. Imposters are said to outnumber true recipients.

Recipients

In total, 3530 medals have been awarded to 3,511 different people. Nineteen men received a second award; 14 of these received two separate Medals for two separate actions, and five received both the Navy and the Army Medals of Honor for the same action. Since the beginning of World War II, 860 Medals of Honor have been awarded, 533 posthumously. In total, 621 had their Medals presented posthumously.

The Army Medal of Honor was first awarded to Private Jacob Parrott during the American Civil War for his role in the Andrews Raid; the most recent Medal was awarded on September 11, 2011 to Dakota Meyer of the Marine Corps for his action in Afghanistan.

While current regulations, beginning in 1918, explicitly state that recipients must be serving in the US Armed Forces at the time of performing a valorous act that warrants the award of the Medal of Honor, exceptions have been made. For example, Mary Walker worked as a military contractor, and Charles Lindbergh, while a reserve member of the US Army Air Corps, received his Medal of Honor as a civilian pilot. In addition, the Medal of Honor was presented to the British Unknown Warrior by General Pershing on October 17, 1921; later the US Unknown Soldier was reciprocally awarded the Victoria Cross, Britain's highest award for gallantry, on November 11, 1921. Apart from these few exceptions, Medals of Honor can only be awarded to members of the US Armed Forces, although being a US citizen is *not* a prerequisite. Sixty-one Canadians serving in the US Armed Forces have been awarded the Medal of Honor, most for actions in the Civil War. Since 1900, only four have been awarded to Canadians. In the Vietnam War, Peter C. Lemon was the only Canadian recipient of the Medal of Honor.

Seventeen retroactive Medals of Honor have been awarded for actions that occurred in previous conflicts, such as the Civil War, World War II, Korea and Vietnam. During the current wars in Iraq and Afghanistan, the Medal of Honor has been awarded 16 times, all but four posthumously, for actions that occurred since the withdrawal of US forces from Vietnam in 1973. The first two were earned by Sergeant First-Class Randy Shughart and Master Sergeant Gary Gordon, who were defending downed Black Hawk helicopter pilot Chief Warrant Officer Michael Durant and his crew during the Battle of Mogadishu in 1993. The others were awarded during the Iraq Campaign. In 2005, a posthumous Medal of Honor awarded to Sergeant First-Class Paul R. Smith for actions in Operation Iraqi Freedom was presented to his survivors. In April 2003, Smith killed over 50 Iraqis near Baghdad International Airport, while providing cover for an aid station full of wounded Americans to evacuate. On January 11, 2007, President George W. Bush awarded Marine Corporal Jason Dunham of Scio, New York, the Medal of Honor posthumously for his bravery in Iraq; he threw himself on a grenade to save his fellow Marines during an action near the Syrian border in April 2004. Three medals of Honor were also awarded to living recipients. They include: Salvatore Giunta, of the Army, Leroy Petry of the Army, and Dakota Meyer of the Marines, Travis Atkins of the Army, and John Canley of the Marines.

Distribution of the Medal of Honor by Conflicts and States

There is not much correlation between the Medal of Honor and the state in which someone lives. The Medal of Honor is a very rare reward for someone to receive, and it seems to be becoming rarer and rarer. For example, over 1500 soldiers and other military personnel (such as nurses and chaplains) earned the Medal of Honor during the Civil War. During World War I, which lasted from 1914 until 1918, there were only 126 people who received the Medal of Honor. In World War II (the US was involved from 1941 until 1945), there were 472 recipients, but in the Vietnam War (which lasted from 1955 until 1975), there were only 261.

In the most recent wars that the United States has been engaged in, the war in Iraq (which began in 2003 and is continuous), and the war in Afghanistan (which began in 2001 and is ongoing), there have only been six and 18, respectively.

Given that so many Medals of Honor were given during the American Civil War to Union forces who fought for the North, the states that have the most Medal of Honor recipients are those that were part of the United States (not including the Southern states that seceded) between 1861 and 1865 and that was actively involved in the Civil War (not all of the country was involved in the war; for example, states on the West Coast were not). After the Civil War, however, there is no correlation between states and Medal of Honor recipients beyond a state's population; more populous states have more recipients than those with lower people.

Figure 1: Medal of Honor Distribution By Conflicts

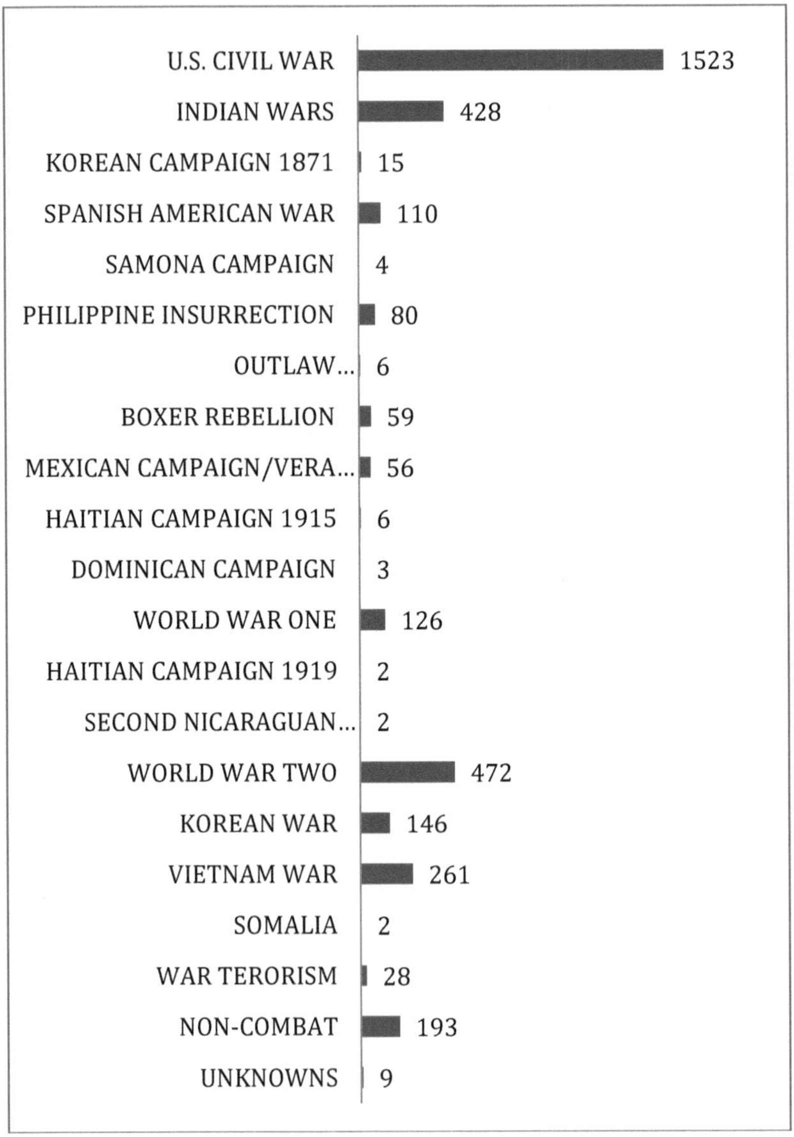

Figure 2: Medal of Honor Distribution By States & Territories

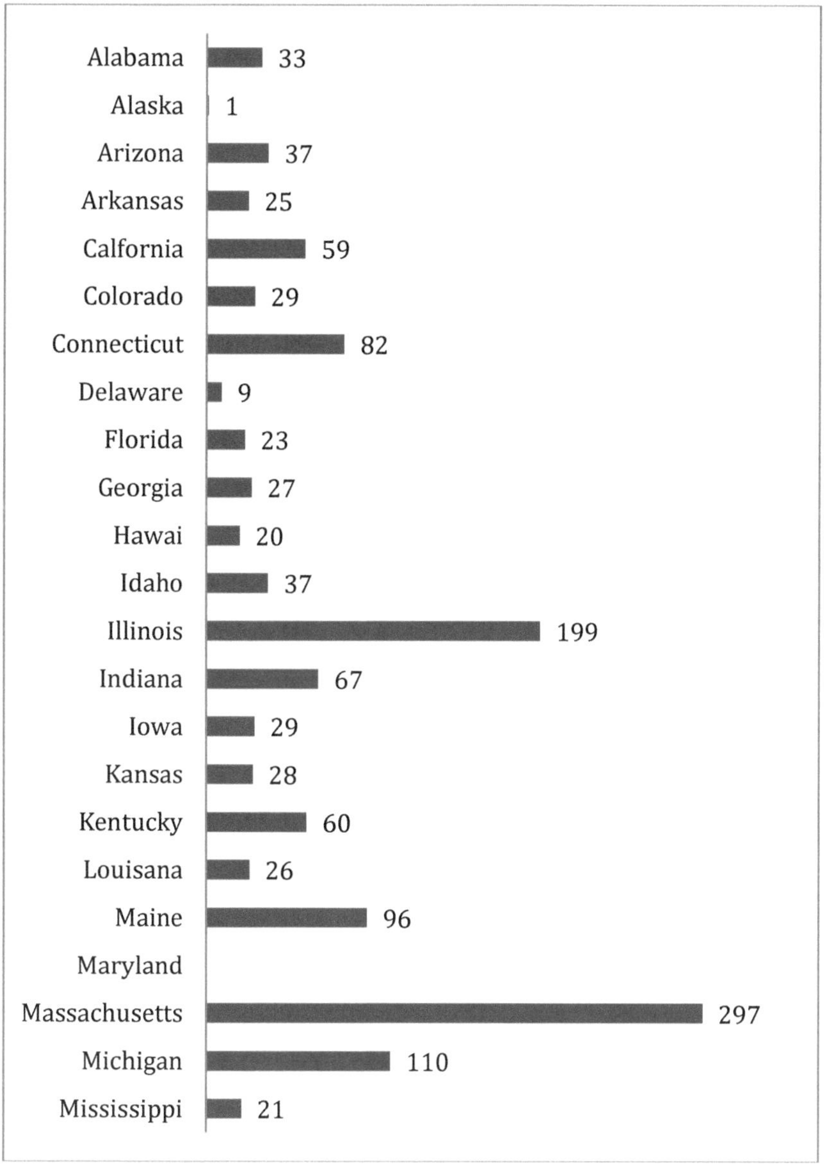

Figure 2: Medal of Honor Distribution By States & Territories

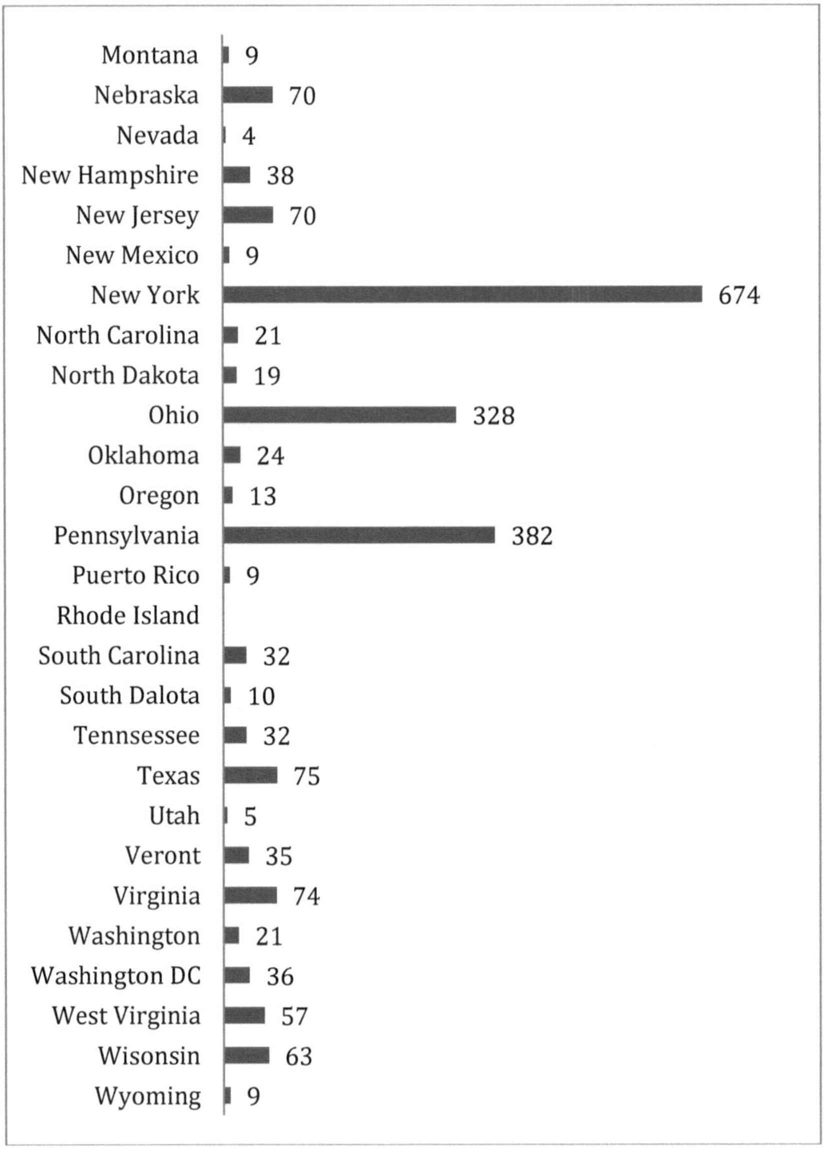

American Civil War Recipients of the Medal of Honor

WILLIAM H. REDDICK

Rank and Organization: Corporal (Highest Rank: Second Lieutenant, Company B, 33rd Ohio Infantry; **Military Service Branch**: U. S. Army; **Conflict**: American Civil War; **Medal of Honor Action Date**: April, 1862; **Medal of Honor Action Place**: Georgia, USA

Citation

One of 19 of 24 men (including two civilians) who, by direction of Gen. Ormsby M. Mitchell, penetrated nearly 200 miles south into enemy territory and captured a railroad train at Big Shanty, Ga., in an attempt to destroy the bridges and track between Chattanooga and Atlanta.

Presentation: March 25, 1863 by Secretary of War Stanton

Accredited to: Portsmouth, Scioto County, Ohio

Born: September 18, 1840, Locust Grove, Adams County, Ohio

Died: November 8, 1903, Letts, Iowa

Buried: Lettsville Cemetery (MH), Letts, Iowa.

Location of Medal: Unknown

Epilogue

Reddick enlisted with the 33rd Ohio Infantry on August 18, 1861. In the spring of 1862, he volunteered to join a group selected from three Ohio regiments by James J. Andrews. Their mission was to steal a train in Georgia and destroy railway facilities. The operation ultimately was unsuccessful, and Reddick was imprisoned in Atlanta, along with thirteen of the other raiders (Andrews and seven others were executed.) He attempted to escape in October of 1862, but was recaptured and sent to Castle Thunder Prison in Richmond, Virginia in December, 1862. After a prisoner exchange in March, 1863, he traveled to Washington, DC where he and five of his fellow raiders received the Medal of Honor. They were the first recipients of the medal.

RICHARD TAYLOR

Rank and Organization: Private, Company E, 18th Indiana Infantry; **Military Service Branch**: U. S. Army; **Conflict**: American Civil War; **Medal of Honor Action Date**: October 19, 1864; **Medal of Honor Action Place**: Cedar Creek, Virginia, USA

Citation

Capture of Flag

Accredited to: Indianapolis, Marion County, Indiana

Born: 1834, Madison County, Alabama

Died: February 23, 1890, Washington, Indiana

Buried: St John's Church (Now Pioneer Park) (MH) Washington, Indiana

Location of Medal: Unknown

Epilogue

Taylor was a Union Army soldier during the American Civil War. He received the Medal of Honor for gallantry during the Battle of Cedar Creek fought near Middletown, Virginia on October 19, 1864. The battle was the decisive engagement of Major General Philip Sheridan's Valley Campaigns of 1864 and was the largest battle fought in the Shenandoah Valley.

Taylor joined the Army from Indiana in August 1861, and mustered out with his regiment in August 1865.

EDWARD J. HOUGHTON

Rank and Organization: Ordinary Seaman, U. S. Picket Boat No. 1; **Military Service Branch**: U. S. Navy; **Conflict**: American Civil War; **Medal of Honor Action Date**: October 27, 1864; **Medal of Honor Action Place**: Plymouth, North Carolina

Citation

Houghton served on board the U.S. Picket Boat No. 1, in action, 27 October 1864, against the Confederate ram, Albemarle, which had resisted repeated attacks by our steamers and had kept a large force of vessels employed in watching her. The picket boat, equipped with a spar torpedo, succeeded in passing the enemy pickets within 20 yards without being discovered and then made for the Albemarle under a full head of steam. Immediately taken under fire by the ram, the small boat plunged on, jumped the log boom which encircled the target, and exploded its torpedo under the port bow of the ram. The picket boat was destroyed by enemy fire and almost the entire crew taken prisoner or lost.

Accredited to: Alabama

Born: 1843, Mobile, Alabama, Mobile County, Alabama

Died: July 16, 1868, Norfolk, Virginia

Buried: Hollywood Cemetery (MH), Brookline, Massachusetts

Epilogue

Houghton was born in Mobile, Alabama in 1843. He served as an Ordinary Seaman in the U.S. Navy during the Civil War. He earned his Medal of Honor on October 27, 1864, whilst serving aboard U.S. Picket Boat No.1. Houghton died in Norfolk, Virginia on July 16, 1868, and is now buried in Hollywood Cemetery, Brookline, Massachusetts.

Spanish American War Recipient of the Medal of Honor

RICHARD PEARSON HOBSON

Rank and Organization: Lieutenant (Highest Rand: Rear Admiral); U.S.S Merrimac; **Military Service Branch**: U. S Navy; **Conflict**: Spanish American War; **Medal of Honor Action Date**: June 3, 1898; **Medal of Honor Action Place**: Santiago de Cuba Harbor Entrance, Cuba.

Citation

In connection with the sinking of the U.S.S. Merrimac at the entrance to the fortified harbor of Santiago de Cuba, 3 June 1898. Despite persistent fire from the enemy fleet and fortifications onshore, Lt. Hobson distinguished himself by extraordinary courage and carried out this operation at the risk of his own personal safety.

Presentation: April 29, 1933 by President Franklin D. Roosevelt.

Accredited to: New York

Born: August 17, 1870, Greensboro, Hale County, Alabama

Died: March 16, 1937, New York, New York

Buried: Arlington National Cemetery (PMH) (6-5014), Arlington, Virginia

Epilogue

Hobson was born at Magnolia Grove in Greensboro in Hale County in the western Black Belt of Alabama. He was the son of Sarah Pearson and James M Hobson. He was the nephew of Richmond Pearson (1852–1923) and the grandson of Justice Richmond Mumford Pearson (1805–1878). He graduated from the U.S. Naval Academy in 1889. He was ostracized by his fellow midshipmen for his total abstinence from alcohol and tobacco. He maintained a superb academic record, graduating first in his class and became the highest ranking midshipman: cadet battalion commander (today's brigade commander). After duty in Chicago, he underwent additional training and was appointed assistant Naval Constructor in 1891. Hobson then served at various navy yards and facilities, including a tour of duty as instructor at the Naval Academy.

Hobson was advanced ten numbers in grade after the war and was promoted to naval constructor with the rank of lieutenant to date from June 23, 1898. Following the end of the war, he helped repair and refit captured Spanish cruisers at Cavite and at various Navy shore stations. In 1899 he became a Veteran Companion of the Pennsylvania Commander of the Military Order of Foreign Wars. He was also a Companion of the New York Commander of the Naval and Military Order of the Spanish War.

As of January 1, 1902 Hobson had been promoted to captain with date of rank of June 23, 1898. After an unsuccessful attempt to retire, he resigned from the Navy in 1903.

In 1933, he was awarded the Medal of Honor for his attempt to block the channel. He was the only Navy officer to receive the Medal of Honor for heroism during the Spanish–American War.

After resigning from active duty in 1903, Hobson remained a staunch supporter of the navy, and, during his subsequent career as a Democratic U.S. Representative from Alabama, 1907–1915, was a firm advocate of naval expansion. He was denied re-nomination in the 1914 Democratic primary, and was notable for being the only congressman from the Deep South to vote for the (failed) women's suffrage bill in the 1915 lame-duck session. In 1934, by special act of the United States Congress, he was advanced to naval constructor with a rank of rear admiral, and placed on the retired list. He died on March 16, 1937, in New York City and is interred in Arlington National Cemetery in Virginia.

In 1905, Hobson married Grizelda Houston Hull, cousin of U.S. Army General Joseph Wheeler, in Tuxedo Park, New York. The couple's son, Richmond P. "Rich" Hobson, Jr., became a rancher in Canada and wrote several popular memoirs of his time there. Hobson's nephew, James Hobson Morrison, Sr., was the Democratic congressman from the Baton Rouge-centered Sixth Congressional District of Louisiana from 1943 to 1967.

One of Hobson's close friends was the inventor Nikola Tesla. Tesla served as the best man in Hobson's wedding. In later years, Hobson was the only person who was able to persuade Tesla to interrupt his intellectual pursuits for a movie gathering.

The destroyer USS *Hobson* (DD-464) was named in his honor. A small town in south Texas was renamed from Castine to Hobson after he spoke there on a railroad tour. A small town in north-central Alabama, Hobson City, Alabama, also took his name.

Interim Period Recipient of the Medal of Honor

GORDON JOHNSTON

Rank and Organization: 1st Lieutenant (Highest Rank: Colonel), U.S. Signal Corps; **Military Service Branch**: U. S. Army; **Conflict**: Philippine Insurrection; **Medal of Honor Action Date**: March 7, 1906; **Medal of Honor Action Place:** Mount Bud Dajo, Jolo, Philippine Island

Citation

Voluntarily took part in and was dangerously wounded during an assault on the enemy's works.

Presentation: The White House, presented by President William H. Taft

Accredited to: Birmingham, Jefferson County, Alabama

Born: May 25, 1874, Charlotte, Mecklenburg County, North Carolina

Died: March 8, 1934, Fort Sam Houston, Texas

Buried: Arlington National Cemetery (MH) (7-10092), Arlington, Virginia

Location of Medal: Princeton University Library, Princeton, New Jersey

Epilogue

Johnston was the son of Confederate General Robert Daniel Johnston, and was born in "the old home" near Birmingham, Alabama. After graduating from Birmingham High School, Johnston enrolled at Princeton University in New Jersey. While attending the university, Johnston played college football as a tackle for two years in 1894 and 1895. Football historian Parke H. Davis described Johnston as a "high spirited, fleetfooted, dashing" football player at Princeton. Johnston became a member of The Tiger Inn. He graduated from Princeton in 1896 and became the head football coach of the University of North Carolina football team in 1896.

In 1896, he also worked in the insurance business as a member of the Johnston & Badham firm in North Carolina. From 1897 to 1898, he was associated with the Buck Johnston Abstract Co. in Memphis, Tennessee.

In 1898, Johnston enlisted in the United States Army to fight in the Spanish–American War, serving in Cuba with the 1st U.S. Volunteer Cavalry Regiment, better known as the Rough Riders. After a brief return to civilian life, he was offered a commission as a second lieutenant in the Army in 1899 on the recommendation of Theodore Roosevelt.

Johnston was posted to the Philippines where he served with the 43rd Infantry Regiment in the Philippine–American War. On February 1, 1900 Johnston's actions as the leader of a small detachment of scouts was to earn him the nation's second highest award for bravery, the Distinguished Service Cross. The citation states: "... *Lieutenant Johnston displayed remarkable gallantry and leadership in charging a greatly superior force of entrenched insurgents in the face of cannon and rifle fire, driving the enemy from their position and capturing the town of Palo."*

Johnston returned to the United States, where he was the honor graduate in 1903 from the U.S. Army's infantry and cavalry school. In 1904, he was married to Anna Julia Johnson in Baltimore, Maryland.

Johnston returned to the Philippines as a first lieutenant in the Signal Corps with the 6th Infantry Regiment. On March 7, 1906, Johnston distinguished himself under heavy fire in the First Battle of Bud Dajo, where he was severely wounded. For his actions in this battle, Johnson received the Medal of Honor.

Johnston went on to lead a distinguished career in the Army. Johnston was to win the Distinguished Service Medal for his work as the chief of staff for the 82nd Infantry Division during the Meuse-Argonne Offensive of World War I. Johnston died in a polo accident at Fort Sam Houston on March 8, 1934.

He was buried at Arlington National Cemetery, in Arlington, Virginia. Camp Gordon Johnston in Carrabelle, Florida was named in his honor.

World War I Recipients of the Medal of Honor

SIDNEY MANNING

Rank and Organization: Corporal, Company G, 167th Infantry, 42nd Division; **Military Service Branch**: U. S. Army; **Conflict**: World War I; **Medal of Honor Action Date**: July 28, 1918; **Medal of Honor Action Place**: Near Breuvannes, France

Presentation: March 17, 1918 by General John J. Pershing near Remagenon, France.

Accredited to: Flomaton, Escambia County, Alabama

Born: July 17, 1892, Butler County, Alabama

Died: December 15, 1960, Flomaton, Alabama

Buried: Little Escombia Baptist Church, Flomaton, Alabama

Citation

When his platoon commander and platoon sergeant had both become casualties soon after the beginning of an assault on strongly fortified heights overlooking the Ourcq River, Cpl. Manning took command of his platoon, which was near the center of the attacking line. Though himself severely wounded he led forward the 35 men remaining in the platoon and finally succeeded in gaining a foothold on the enemy's position, during which time he had received more wounds and all but seven of his men had fallen. Directing the consolidation of the position, he held off a large body of the enemy only 50 yards away by fire from his automatic rifle. He declined to take cover until his line had been entirely consolidated with the line of the platoon on the front when he dragged himself to shelter, suffering from nine wounds in all parts of the body.

Epilogue

Manning was born in Butler County, Alabama on July 17, 1892, the son of Barney Austin Manning and Sarah Elizabeth (Vickery) Manning. Manning's siblings included three brothers and a sister. He was raised on a farm in Flomaton, Alabama, and educated in Flomaton. When he registered for the World War I draft in June 1917, Manning stated that farming was his occupation. When he joined the military, he was assigned to the Alabama National Guard's Company G, 4th Alabama Infantry. In August, the 4th Alabama entered federal service as the 167th Infantry Regiment, a unit of the 42nd Division.

In 1921, Manning accepted a U.S. government invitation to attend the inaugural Veterans Day wreath lying at the Tomb of the Unknown Soldier in Washington, D.C. In 1935, the Alabama legislature enacted legislation that provided financial support to Manning for the purchase of a farm and equipment to operate it. Manning never fully recovered from his wounds; in addition to receiving a disability pension, the condition of his health required him to frequently visit Veterans Administration hospitals for treatment and evaluation. His health continued to decline, and in his later years, Manning was completely bedridden.

Manning died in Montgomery, Alabama on December 15, 1960. He was buried at Little Escambia Cemetery in Flomaton.

In 1976, the town of Flomaton dedicated a granite memorial in Manning's honor. Originally placed at Lion's Park, it was later moved to the Flomaton Welcome Center. The welcome center is located on Sidney E. Manning Boulevard, which is also named for Manning. In 1962, a plaque honoring Manning was placed at the Alabama Department of Archives and History in Montgomery.

In 1919, Manning married Mamie Knowles (1899-1977). They were the parents of two children, son Sidney Hill Manning (1921-1995) and a daughter who died young.

OSMOND KELLY INGRAM

POSTHUMOUS AWARD

Rank and Organization: Gunner's Mate First Class, U.S.S. Cassin; **Military Service Branch**: U. S. Navy; **Conflict**: World War I; **Medal of Honor Action Date**: October 15, 1917; **Medal of Honor**

Action Place: At Sea 20 miles south of Mind Head Lighthouse, Monagoush, County Waterford, Ireland.

Accredited to: Pratt City, Jefferson County, Alabama

Born: August 4, 1887, Onneonta County, Alabama

Died: October 15, 1917 at sea

Buried: A.B.M.C. Wall of the Missing, Brookwood, Surrey, England

Location of Medal: Birmingham Museum of Art, Birmingham, Alabama

Citation

For extraordinary heroism in the presence of the enemy on the occasion of the torpedoing of the Cassin, 15 October 1917. While the Cassin was searching for the submarine, Ingram sighted the torpedo coming, and realizing that it might strike the ship aft in the vicinity of the depth charges, ran aft with the intention of releasing the depth charges before the torpedo could reach the Cassin. The torpedo struck the ship before he could accomplish his purpose, and Ingram was killed by the explosion. The depth charges exploded immediately afterward. His life was sacrificed in an attempt to save the ship and his shipmates, as the damage to the ship would have been much less if he had been able to release the depth charges.

Epilogue

Born to Robert L. Ingram and his wife Naomi Elizabeth Lea in Oneonta, Alabama, Ingram entered the Navy November 24, 1903. His ship, USS *Cassin*, was attacked by the German submarine *U-61* off Ireland on October 15, 1917. Gunner's Mate First Class Ingram spotted the approaching torpedo, realized it would strike close by the ship's depth charges, thus dooming the ship, and rushed to jettison the ammunition. He was blown overboard when the torpedo struck, thus becoming the United States' Navy's first enlisted man killed in action in World War I as he attempted to save his ship and shipmates. He posthumously received the Medal of Honor for his actions on that day.

The main flagpole and colors at the former Naval Training Center, San Diego are named Ingram Plaza in his honor

USS *Osmond Ingram* (DD-255), a *Clemson*-class destroyer, served in the United States Navy during World War II. *Osmond Ingram* was decommissioned at Philadelphia on January 8, 1946, and was struck from the Navy List on January 21, 1946. It was sold for scrapping to Hugo Neu, June 17, 1946.

Kelly Ingram Park is also named in his honor. It is a 4-acre (1.6 ha) park located in Birmingham, Alabama.

GM1 O. K. Ingram aboard USS *Cassin* on October 15, 1917

There is a Veterans of Foreign Wars Post Named after him in Birmingham, AL Post 668.

World War II Recipients of the Medal of Honor

GEORGE WATSON

POSTHUMOUS AWARD

Rank and Organization: Private, 2nd Battalion, 29th Quartermaster Regiment; **Military Service Branch**: U. S. Army; **Conflict**: World War II; **Medal of Honor Action Date**: March 8, 1943; **Medal of Honor Action Place**: Near Porlock Harbor, New Guinea.

Accredited to: Birmingham, Jefferson County, Alabama

Born: March 24, 1914, Birmingham, Jefferson County, Alabama

Died: March 8, 1943, Porlock Harbor, New Guinea

Buried: Buried at sea

Location of Medal: U. S. Army Quartermaster Museum, Fort lee, Virginia.

Citation

For extraordinary heroism in action on 8 March 1943. Private Watson was on board a ship which was attacked and hit by enemy bombers. When the ship was abandoned, Private Watson, instead of seeking to save himself, remained in the water assisting several soldiers who could not swim to reach the safety of the raft. This heroic action, which subsequently cost him his life, resulted in the saving of several of his comrades. Weakened by his exertions, he was dragged down by the suction of the sinking ship and was drowned. Private Watson's extraordinarily valorous actions, daring leadership, and self-sacrificing devotion to his fellow-man exemplify the finest traditions of military service.

Epilogue

Born and raised in Birmingham, Alabama, Watson was 29 when he died, pulled down by a sinking ship. When the Medal was finally awarded to recognize his actions some fifty years later, there was no next of kin to receive it. The Medal was instead given to the Army's Sergeant Major for safe-keeping. Watson's body was never recovered.[7]

PAUL L. BOLDEN

Rank and Organization: Staff Sergeant (Highest Rank: Master Sergeant, Company I, 120th Infantry, 30the Division; Military **Service Branch**: U. S. Army; **Conflict**: World War II; **Medal of Honor Action Date**: December 23, 1944; **Medal of Honor Action Place**: Petit-Coo, Belgium

Presentation: August 23, 1945, at The White House by President Harry S Truman

Accredited to: Madison, Madison County, Alabama

Born: June 15, 1922, Hobbes Island, Madison County, Alabama

Died: May 21, 1971, Owens Crossroads, Alabama

Buried: Moons Cemetery (MH), Owens Crosswoads, Alabama

Location of Medal: Birmingham Museum of Art, Birmingham, Alabama

Citation

He voluntarily attacked a formidable enemy strongpoint in Petit-Coo, Belgium, on 23 December 1944, when his company was pinned down by extremely heavy automatic and small-arms fire coming from a house 200 yards to the front. Mortar and tank artillery shells pounded the unit, when SSgt. Bolden and a comrade, on their own initiative, moved forward in a hail of bullets to eliminate the ever-increasing fire from the German position. Crawling ahead to close with what they knew was a powerfully armed, vastly superior force, the pair reached the house and took up assault positions, SSgt. Bolden under a window, his comrade across the street where he could deliver covering fire. In rapid succession, SSgt. Bolden hurled a fragmentation grenade and a white phosphorous grenade into the building; and then, fully realizing that he faced tremendous odds, rushed to the door, threw it open, and fired into 35 SS troopers who were trying to reorganize themselves after the havoc wrought by the grenades. Twenty Germans died under fire of his submachine gun before he was struck in the shoulder, chest, and stomach by part of a burst which killed his comrade across the street. He withdrew from the house, waiting for the surviving Germans to come out and surrender. When none appeared in the doorway, he summoned his ebbing strength, overcame the extreme pain he suffered, and boldly walked back into the house, firing as he went. He had killed the remaining 15 enemy soldiers when his ammunition ran out. SSgt. Bolden's heroic advance against great odds, his fearless assault, and his magnificent display of courage in reentering the building where he had been severely wounded cleared the path for his company and insured the success of its mission.

Epilogue

Bolden joined the Army from Madison, Alabama in October 1942, and by December 23, 1944, was serving as a Staff Sergeant in Company I, 120th Infantry Regiment, 30th Infantry Division. On that day, at Petit-Coo, Belgium, he and Technical Sergeant Russell N. Snoad advanced on a German-held house. While his comrade provided covering fire from across the street, Bolden tossed grenades through a window, rushed to the door, and began firing. Wounded by the greatly superior number of German soldiers inside, he retreated from the house. Realizing that the Germans would not surrender, he returned to the house despite his serious wounds and killed the remaining soldiers. For these actions, he was awarded the Medal of Honor eight months later, on August 23, 1945.

Bolden reached the rank of Master Sergeant before leaving the Army. He died at age 56 and was buried in Moon Cemetery, Owens Cross Roads, Alabama.

An 18-mile stretch of Alabama State Route 53, from Research Park Boulevard in Huntsville to just south of Ardmore, is designated as the "Paul Luther Bolden Memorial Highway" in his honor.

CECIL H. BOLTON

Rank and Organization: First Lieutenant, Company E, 413th Infantry, 104th Infantry Division; **Military Service Branch**: U. S. Army; **Conflict**: World War II; **Medal of Honor Action Date**: November 2, 1944; **Medal of Honor Action Place**: Mark River, Holland

Presentation: August 23, 1945, at The White House by President Harry S Truman

Accredited to: Huntsville, Madison County, Alabama

Born: October 2, 1908, Crawfordville, Wakulla, Alabama

Died: January 22, 1965, San Antonio, Texas

Buried: Fort Sam Houston National Cemetery (MH) (P-C-22-J), San Antonio Texas

Citation

As leader of the weapons platoon of Company E, 413th Infantry, on the night of 2 November 1944, he fought gallantly in a pitched battle which followed the crossing of the Mark River in Holland. When two machine guns pinned down his company, he tried to eliminate, with mortar fire, their grazing fire which was inflicting serious casualties and preventing the company's advance from an area rocked by artillery shelling. In the moonlight it was impossible for him to locate accurately the enemy's camouflaged positions; but he continued to direct fire until wounded severely in the legs and rendered unconscious by a German shell. When he recovered consciousness he instructed his unit and then crawled to the forward rifle platoon positions. Taking a two-man bazooka team on his voluntary mission, he advanced chest-deep in chilling water along a canal toward one enemy machine gun. While the bazooka team covered him, he approached alone to within 15 yards of the hostile emplacement in a house. He charged the remaining distance and killed the two gunners with hand grenades. Returning to his men he led them through intense fire over open ground to assault the second German machine gun. An enemy sniper who tried to block the way was dispatched, and the trio pressed on. When discovered by the machine-gun crew and subjected to direct fire, 1st Lt. Bolton killed one of the three gunners with carbine fire, and his two comrades shot the others. Continuing to disregard his wounds, he led the bazooka team toward an 88-mm artillery piece which was having telling effect on the American ranks, and approached once more through icy canal water until he could dimly make out the gun's silhouette. Under his fire direction, the two soldiers knocked out the enemy weapon with rockets. On the way back to his own lines he was again wounded. To prevent his men being longer subjected to deadly fire, he refused aid and ordered them back to safety, painfully crawling after them until he reached his lines, where he collapsed. First Lt. Bolton's heroic assaults in the face of vicious fire, his inspiring leadership, and continued aggressiveness even through suffering from serious wounds contributed in large measure to overcoming strong enemy resistance and made it possible for his battalion to reach its objective.

Epilogue

Bolton joined the Army from Ft McClellan, near Anniston, Alabama on the 27th of July, 1942, and by November 2, 1944 was serving as a first lieutenant in Company E, 413th Infantry Regiment, 104th Infantry Division. On that day, near the Mark river in North Brabant, the Netherlands, he was seriously wounded in the legs by a German artillery shell. Despite these wounds, he took two men and led them in a successful assault against three German positions which were firing on his company. Wounded a second time, he ordered his two companions to leave him behind and head for the safety of the American lines. He then crawled the rest of the way back to his company. For these actions, he was awarded the Medal of Honor ten months later, on September 1, 1945.

Bolton reached the rank of colonel before leaving the Army. He died at age 56 and was buried in Fort Sam Houston National Cemetery, San Antonio, Texas.

JOHN D. NEW

POSTHUMOUS AWARD

Rank and Organization: Private First Class, 2nd Battalion, 7th Marine Regiment, 1st Marine Division; **Military Service Branch**: U. S. Maine Corps; **Conflict**: World War II; **Medal of Honor Action Date**: September 25, 1944; **Medal of Honor Action Place**: Peleliu Island, Palau Islands

Presentation: June 23, 1945, Naval Air Station, Pensacola, Florida, presented by Rear Admiral C.A. Pownall to his father

Accredited to: Alabama

Born: August 12, 1924, Mobile, Mobile County, Alabama

Died: September 25, 1944, Peleliu Island, Palau Islands

Buried: Mobile National Cemetery (MH) (7-2147), Mobile, Alabama

Citation

For conspicuous gallantry and intrepidity at the risk of his life above and beyond the call of duty while serving with the 2d Battalion, 7th Marines, 1st Marine Division, in action against enemy Japanese forces on Peleliu Island, Palau Group, 25 September 1944. When a Japanese soldier emerged from a cave in a cliff directly below an observation post and suddenly hurled a grenade into the position from which two of our men were directing mortar fire against enemy emplacements, Pfc. New instantly perceived the dire peril to the other marines and, with utter disregard for his own safety, unhesitatingly flung himself upon the grenade and absorbed the full impact of the explosion, thus saving the lives of the two observers. Pfc. New's great personal valor and selfless conduct in the face of almost certain death reflect the highest credit upon himself and the U.S. Naval Service. He gallantly gave his life for his country.

Epilogue

New was born to Jesse James New and Mary Magdalene Gambrel in Mobile, Alabama, and attended St. Joseph's Parochial School and the public schools of that city.

He enlisted in the United States Marine Corps on December 8, 1941, the first Mobile man to enlist after Pearl Harbor. On December 12, 1941, he joined the 4th Recruit Battalion at the Marine Corps Recruit Depot Parris Island, South Carolina, and following recruit training served at Quantico, Virginia.

Grave of PFC John D. New at the Mobile (Alabama) National Cemetery

In September 1942, Private New joined the 4th Replacement Battalion, Fleet Marine Force, Training Center, at San Diego, California. He sailed for duty overseas the following month, and took part in the Guadalcanal campaign as a member of 2nd Battalion, 7th Marines, 1st Marine Division. He shared in the Presidential Unit Citation awarded the Division on Guadalcanal. He later took part in the seizure and occupation of Cape Gloucester, New Britain, and in the defense of the airfield in the vicinity of Cape Gloucester.

Promoted to private first class in April 1944, he moved from Cape Gloucester to Pavuvu, Russell Islands, to Guadalcanal, and on September 15, 1944 sailed for Peleliu in the Palau Islands. Nine days later, he was killed in action on Peleliu when he flung himself on an enemy hand grenade to save the lives of two comrades.

The Medal of Honor, awarded PFC New posthumously, was presented to his father.

Initially buried in the U.S. Armed Forces Cemetery on Peleliu, PFC New's remains were reinterred in the Mobile National Cemetery, Mobile, Alabama, in 1948. His grave can be found in section 7-2147.

ROSS F. GRAY

POSTHUMOUS AWARD

Rank and Organization: Sergeant, Company A, 1st battalion, 25th marines, 4th Marine Division; **Military Service** Branch: U. S. Maine Corps; Conflict: World War II; **Medal of Honor Action Date**: February 21, 1945; **Medal of Honor Action Place**: Iwo Jima Island

Presentation: April 16, 1946. Presented to his mother by Admiral A. S. Merrill at Centerville High School, Centerville, Alabama

Accredited to: Alabama

Born: August 1, 1920, Marvel Valley, Bibb County, Alabama

Died: February 27, 1945, Iwo Jima Island

Buried: Ada Chapel Cemetery (MH), West Blocton, Alabama

Citation

For conspicuous gallantry and intrepidity at the risk of his life above and beyond the call of duty as a platoon sergeant attached to Company A, 1st Battalion, 25th Marines, 4th Marine Division, in action against enemy Japanese forces on Iwo Jima, Volcano Islands, 21 February 1945. Shrewdly gaging the tactical situation when his platoon was held up by a sudden barrage of hostile grenades while advancing toward the high ground northeast of Airfield No. 1, Sgt. Gray promptly organized the withdrawal of his men from enemy grenade range, quickly moved forward alone to reconnoiter, and discovered a heavily mined area extending along the front of a strong network of emplacements joined by covered trenches. Although assailed by furious gunfire, he cleared a path leading through the minefield to one of the fortifications, then returned to the platoon position and, informing his leader of the serious situation, volunteered to initiate an attack under cover of three fellow marines. Alone and unarmed but carrying a huge satchel charge, he crept up on the Japanese emplacement, boldly hurled the short-fused explosive, and sealed the entrance. Instantly taken under machine-gun fire from a second entrance to the same position, he unhesitatingly braved the increasingly vicious fusillades to crawl back for another charge, returned to his objective, and blasted the second opening, thereby demolishing the position. Repeatedly covering the ground between the savagely defended enemy fortifications and his platoon area, he systematically approached, attacked, and withdrew under blanketing fire to destroy a total of six Japanese positions, more than 25 troops, and a quantity of vital ordnance gear and ammunition. Stouthearted and indomitable, Sgt. Gray had singlehandedly overcome a strong enemy garrison and had completely disarmed a large minefield before finally rejoining his unit. By his great personal valor, daring tactics, and tenacious perseverance in the face of extreme peril, he had contributed materially to the fulfillment of his company's mission. His gallant conduct throughout enhanced and sustained the highest traditions of the U.S. Naval Service.

Epilogue

Gray was born in Marvel Valley, Alabama. He attended the elementary schools of Bibb County and went on to Centreville High School, which he left in 1939 after three years in attendance.

He went to work for his father as a carpenter. He had already worked at that trade part-time for three years and now worked another three years before he enlisted in the Marine Corps Reserve. He played football and basketball at Centreville High and enjoyed hunting and fishing. He was a devout Protestant; his comrades in the Marine Corps called him "The Deacon."

Gray enlisted in the Marine Corps Reserve in Birmingham, Alabama on July 22, 1942, and was assigned to active duty the same day. After receiving his recruit training at Parris Island, South Carolina, he went to New River, North Carolina, and in September joined the 23rd Marines, 4th Marine Division. Promoted to private first class in April 1943, he was transferred to Company A, 1st Battalion 25th Marines, a month later.

Gray left for overseas duty on January 13, 1944, and landed at Kwajalein in the Marshall Islands where he took part in the Roi-Namur campaign. He was made an engineering corporal in March and in June made another assault landing — this time at Saipan. At the conclusion of the fighting at Saipan, Cpl Gray took part in the landing on Tinian Island, also in the Marianas.

Promoted to sergeant in August, he attended the 4th Marine Division Mine and Booby Trap School, upon completion of which he was rated qualified to instruct troops in the laying of mine fields; the reconnaissance of enemy minefields, day and night; the location, neutralization, disarming, and removal of mines; the neutralization of booby-trapped mines; and the day and night clearance of lanes through minefields. Examined and found qualified for promotion to the rank of staff sergeant, Sgt Gray, due to the lack of openings for that rate in his organization, was never promoted to the third pay grade.

On February 21, 1945, two days after D-Day on Iwo Jima, Sgt Gray was acting platoon sergeant of one of Company A's platoons which had been held up by a sudden barrage of Japanese hand grenades in the area northeast of Airfield No. 1. Promptly withdrawing his platoon out of range of the grenades, he moved forward to reconnoiter. He ascertained that the advance was held up by a series of enemy emplacements connected by covered communication trenches and fronted by a mine field.

Through a hail of enemy small arms fire, Sgt Gray cleared a path through the mine field up to the mouth of one of the fortifications, then returned to his own lines, where with three volunteers, he went back to the battalion dump and acquired twelve satchel charges. Placing these in a defiladed area within his platoon that was protected from immediate enemy fire, he took one weighing twenty-four pounds. Under covering fire from the three volunteers, Sgt Gray advanced up the path he had cleared and threw the charge into the enemy position in order to neutralize it.

Immediately brought under fire from a machine gun in another opening of the same position, Sgt Gray returned to the defiladed spot, obtained another charge, returned to the position and this time completely destroyed it. Spotting another emplacement, he went through the mine field for the seventh and eighth time to get another charge and destroy another enemy stronghold.

He continued this one-man attack, all the time under heavy small arms fire and grenade barrage, until he had destroyed six enemy positions. At one time a Japanese grenade landed so close to him that it blew his helmet off. Throughout, he went unarmed so that he could more easily carry the charges and accessories.

When he had eliminated the six positions, Sgt Gray disarmed the whole mine field before returning to his platoon.

Although he remained unscratched through his twelve trips back and forth among enemy mines, constantly under heavy fire, Sgt Gray was killed six days later, on February 27, 1945, by an enemy shell which inflicted fatal wounds in his legs.

For his personal valor, daring tactics, and tenacious perseverance in the face of extreme peril on February 21, Sgt Gray was posthumously awarded the Medal of Honor by President Harry S. Truman.

The coveted award was presented to the hero's father by Rear Admiral A. S. Merrill, United States Navy, then Commandant of the Eighth Naval District, at the football field at Centreville High School in the presence of the Governor of the State of Alabama, Chauncey Sparks, on April 16, 1946.

Sergeant Gray was initially buried in the 4th Marine Division Cemetery on Iwo Jima, but later his remains were returned to the United States for private burial in Woodstock, Alabama.

HOWARD W. GILMORE

POSTHUMOUS AWARD

Rank and Organization: Commander, U.S.S. Growley; **Military Service Branch**: U. S. Navy; **Conflict**: World War II; **Medal of Honor Action Date**: January 10-February 7, 1945; **Medal of Honor Action Place**: Southwest Pacific

Presentation: July 13, 1945, New Orleans, Louisiana. Presented by Rear Admiral A.C. Bennett to his widow.

Accredited to: Louisiana

Born: September 29, 1902, Selma, Dallas County, Alabama

Died: February 7, 1943 at sea

Buried: A.B.M.C. Manila Cemetery (Wall of the Missing); Remains not recovered, Manila, Philippine Island

Location of Medal: National Museum of the Pacific War, Fredericksburg, Texas

Citation

For distinguished gallantry and valor above and beyond the call of duty as commanding officer of the U.S.S. Growler during her fourth war patrol in the southwest Pacific from 10 January to 7 February 1943. Boldly striking at the enemy in spite of continuous hostile air and anti-submarine patrols, Comdr. Gilmore sank one Japanese freighter and damaged another by torpedo fire, successfully evading severe depth charges following each attack. In the darkness of night on 7 February, an enemy gunboat closed range and prepared to ram the Growler. Comdr. Gilmore daringly maneuvered to avoid the crash and rammed the attacker instead, ripping into her port side at 17 knots and bursting wide her plates. In the terrific fire of the sinking gunboat's heavy machine guns, Comdr. Gilmore calmly gave the order to clear the bridge, and refusing safety for himself, remained on deck while his men preceded him below. Struck down by the fusillade of bullets and having done his utmost against the enemy, in his final living moments, Comdr. Gilmore gave his last order to the officer of the deck, "Take her down." The Growler dived; seriously damaged but under control, she was brought safely to port by her well-trained crew inspired by the courageous fighting spirit of their dead captain.

Epilogue

Gilmore was born in Selma, Alabama, September 29, 1902 and enlisted in the Navy November 15, 1920. In 1922 he was appointed to the United States Naval Academy by competitive examination. Standing 34th of a class of 436, Gilmore was commissioned in 1926 and reported to the battleship USS *Mississippi* (BB-41). Gilmore underwent submarine training during 1930 and in the years that followed served in various submarines and at stations ashore.

Gilmore served as the executive officer of the submarine USS *Shark* (SS-174) (USS Shark), and in a near-fatal incident during the submarine's shakedown cruise, narrowly survived an assault by a group of thugs in Panama, who cut his throat during an excursion ashore. He had to deal with several other instances of tragedy in his life, including the death of his first wife from disease, and at the time of his Medal of Honor action his second wife was still in a coma from a fall she had taken down a flight of stairs. In 1941, he assumed his first command, USS *Shark* (SS-174), only to be transferred the day after the attack on Pearl Harbor to command the still-unfinished submarine USS *Growler* (SS-215).

Gilmore commanded his submarine skillfully during four Pacific War patrols. During his first, on 5 July 1942 *Growler* attacked three enemy destroyers off Kiska, sinking one and severely damaging the other two, while narrowly avoiding two torpedoes fired in return, for which Gilmore received the Navy Cross.

During his second patrol, *Growler* sank four merchant ships totaling 15,000 tons in the East China Sea near Formosa. Gilmore was awarded a gold star in lieu of a second Navy Cross.

During October 1942, *Growler* patrolled off Truk in the Caroline Islands as part of a repositioning of submarine assets on the way to Brisbane, Australia. No significant action occurred.

The submarine continued to take a heavy toll on shipping on its fourth war patrol, and during the night of 6–7 February 1943, it approached a convoy stealthily for a surface attack. Suddenly a convoy escort named *Hayasaki* closed and prepared to ram. As the small ship charged out of the darkness, Gilmore sounded the collision alarm and shouted, "Left full rudder!" — to no avail. Perhaps inadvertently, *Growler* hit the Japanese adversary amidships at 17 knots (31 km/h), heeling the submarine 50 degrees, bending 18 feet of its bow sideways to port, and disabling the forward torpedo tubes.

Simultaneously, the Japanese crew began a burst of machine gun fire at *Growler's* bridge, killing the junior officer of the deck and a lookout, while wounding Gilmore himself and two other men. "Clear the bridge!" Gilmore ordered as he struggled to hang on to a frame. As the rest of the bridge party dropped down the hatch into the conning tower, the executive officer, Lieutenant Commander Arnold Schade — shaken by the impact and dazed by his own fall into the control room — waited expectantly for his captain to appear. Instead from above came the shouted command, "Take her down!" Realizing that he could not get below in time if the ship were to escape, Gilmore chose to make the supreme sacrifice for his shipmates. Schade hesitated briefly — then obeyed his captain's last order and submerged the crippled ship.

Surfacing some time later in hope of reattacking the *Hayasaki*, Schade found the seas empty. The Japanese ship had, in fact, survived the encounter, but there was no sign of Gilmore, who apparently had drifted away during the night. Schade and *Growler's* crew managed to control the ship's flooding and voyaged back to Brisbane on February 17.

WILLIAM D. WATSON

Rank and Organization: Private (Highest Rank: Staff Sergeant), 2nd Squad, 3rd Platoon, Company G, 2nd Battalion, 9th Marine Division; **Military Service Branch**: U. S. Marine Corps; **Conflict**: World War II; **Medal of Honor Action Date**: February 26, -27, 1945; **Medal of Honor Action Place**: Iwo Jima Island

Presentation: October 5, 1945 at The White House, presented by President Harry S. Truman

Accredited to: Earle, Crittenden County, Arkansas

Born: February 18, 1922, Tuscumbia, Colbert County, Alabama

Died: December 19, 1984, Russellville, Arkansas

Buried: Russell Cemetery (MH), Ozone, Arkansas

Location of Medal: National Museum of the Pacific War, Fredericksburg, Texas

Citation

For conspicuous gallantry and intrepidity at the risk of his life above and beyond the call of duty as automatic rifleman serving with the 2d Battalion, 9th Marines, 3d Marine Division, during action against enemy Japanese forces on Iwo Jima, Volcano Islands, 26 and 27 February 1945. With his squad abruptly halted by intense fire from enemy fortifications in the high rocky ridges and crags commanding the line of advance, Pvt. Watson boldly rushed one pillbox and fired into the embrasure with his weapon, keeping the enemy pinned down singlehandedly until he was in a position to hurl in a grenade and then running to the rear of the emplacement to destroy the retreating Japanese and enable his platoon to take its objective. Again pinned down at the foot of a small hill, he dauntlessly scaled the jagged incline under fierce mortar and machine-gun barrages and, with his assistant BAR man, charged the crest of the hill, firing from his hip. Fighting furiously against Japanese troops attacking with grenades and knee mortars from the reverse slope, he stood fearlessly erect in his exposed position to cover the hostile entrenchments and held the hill under savage fire for 15 minutes, killing 60 Japanese before his ammunition was exhausted and his platoon was able to join him. His courageous initiative and valiant fighting spirit against devastating odds were directly responsible for the continued advance of his platoon, and his inspiring leadership throughout this bitterly fought action reflects the highest credit upon Pvt. Watson and the U.S. Naval Service.

Epilogue

Watson was born on February 18, 1922, to Charles Watson and Ada Watson in Tuscumbia, Alabama. He is one of twelve kids; eight brothers, three sisters. Before his enlistment in Little Rock, Arkansas, on August 6, 1942, he worked on his father's farm and completed seven years of grade school. Watson received his basic training at the Marine Corps Recruit Depot San Diego, California, and was deployed overseas on January 24, 1943.

Serving as an automatic rifleman with Company G, 2nd Battalion, 9th Marines, 3rd Marine Division during the bitter fighting on Iwo Jima, Watson earned the Medal of Honor for heroism during February 26–27, 1945, when he single-handedly killed more than 90 Japanese and enabled his pinned-down platoon to continue the advance. In the attack, he was shot seven times and was hit in the shoulder by mortar fragments. He was evacuated from Iwo Jima after he suffered a gunshot wound to the neck on March 2, 1945. He previously saw action at Bougainville, Guadalcanal, and Guam.

Following his discharge from the Marine Corps, he enlisted in the United States Army Air Force, and then the United States Army as a private, working as a mess hall cook. In 1963, he was arrested in Marion, Arkansas and charged with desertion after being reported missing from his post in Fort Rucker, Alabama in October 1962. When asked about his leaving, Watson stated he just got "fed up" with the Army and left. He eventually reached the rank of Staff Sergeant and finally retired from the military in 1966.

He was married to wife Patricia, whom he had two children with.

CHARLES W. DAVIS

Rank and Organization: Captain (Highest Rank: Major); 2nd Battalion, 27th Infantry Regiment, 25th Infantry Division; **Military Service Branch**: U. S. Army; **Conflict**: World War II; **Medal of Honor Action Date**: January 12, 1943; **Medal of Honor Action Place**: Guadalcanal, Solomon Islands

Presentation: July 30, 1943 Guadalcanal, Solomon Island. Presented by Lt. Gen. Millard F. Harmon.

Accredited to: Montgomery, Montgomery County, Alabama

Born: February 1, 1917, Gordo, Pickens County, Alabama

Died: January 18, 19991, San Francisco, California

Buried: Arlington National Cemetery (MH) (7-A-170), Arlington, Virginia

Citation

For distinguishing himself conspicuously by gallantry and intrepidity at the risk of his life above and beyond the call of duty in action with the enemy on Guadalcanal Island. On 12 January 1943, Maj. Davis (then Capt.), executive officer of an infantry battalion, volunteered to carry instructions to the leading companies of his battalion which had been caught in crossfire from Japanese machine guns. With complete disregard for his own safety, he made his way to the trapped units, delivered the instructions, supervised their execution, and remained overnight in this exposed position. On the following day, Maj. Davis again volunteered to lead an assault on the Japanese position which was holding up the advance. When his rifle jammed at its first shot, he drew his pistol and, waving his men on, led the assault over the top of the hill. Electrified by this action, another body of soldiers followed and seized the hill. The capture of this position broke Japanese resistance and the battalion was then able to proceed and secure the corps' objective. The courage and leadership displayed by Maj. Davis inspired the entire battalion and unquestionably led to the success of its attack.

Epilogue

Born February 21, 1917, Davis joined the Army from Montgomery, Alabama, and by January 12, 1943 was serving as a captain in the 27th Infantry Regiment, 25th Infantry Division. On that day, on the island of Guadalcanal during the Battle of Mount Austen, the Galloping Horse, and the Sea Horse, he volunteered to carry messages to several companies which were pinned down by Japanese fire. He stayed with the companies overnight. A knoll on the south edge of the ridge (the "horse's neck") leading to Hill 53 was the fulcrum of the Japanese defenses. The knoll contained several machine-gun and mortar positions which had effectively held off the American attacks across the ridge. As 2nd Battalion's executive officer Captain Davis volunteered to lead four other men against the knoll. Crawling on their bellies, Davis and his party crept to within 10 yd (9.1 m) of the enemy position. The Japanese defenders threw two grenades at them, but the grenades failed to explode. Davis and his men threw eight grenades at the Japanese, destroying several of their positions. Davis then stood up, and while shooting his rifle, then pistol with one hand, waved his men forward with the other as he advanced further onto the knoll. Davis and his men then killed or chased away the rest of the Japanese on the knoll. Silhouetted against the sky during the action, Davis was visible to the Americans all up and down the ridge. Inspired by his actions, plus replenished with water by a sudden thunderstorm, the American troops "came to life" and quickly assaulted and captured Hill 53 by noon. The Americans counted the bodies of 170 Japanese soldiers on and around the Galloping Horse. The Americans suffered fewer than 100 killed.

On the right, Major General J. L. Collins, commander of the 25th Division and, on the left, Major Charles W. Davis, commanding the 3rd Battalion, 27th Infantry Regiment confer on New Georgia, August 14, 1943.

Davis reached the rank of colonel and served in the Korean and Vietnam Wars before leaving the Army. He died at age 73 and is buried in Arlington National Cemetery, Arlington County, Virginia.

WILLIAM R. LAWLEY JR.

Rank and Organization: 1st Lieutenant (Highest Rank: Colonel), 364th Bombardment Squadron, 395th Bombardment Group, 40th Combat Wing, 1s Air Division, 8th Air Force; **Military Service Branch**: U. S. Army Air Corps; **Conflict**: World War II; **Medal of Honor Action Date**: February 20, 1944; **Medal of Honor Action Place**: Over Leipzig, Germany

Presentation: August 8, 1944, Headquarters U. S. Army Air Forces Europe, High Wycomb, England. Presented by Lt. Gen. Carl A. Spaatz

Accredited to: Birmingham, Jefferson County, Alabama

Born: August 23, 1920, Leeds, Jefferson County, Alabama

Died: May 29, 1999, Montgomery, Alabama

Buried: Greenwood Cemetery (MH) (12-B-10-C-1), Montgomery, Alabama

Citation

For conspicuous gallantry and intrepidity in action above and beyond the call of duty, 20 February 1944, while serving as pilot of a B-17 aircraft on a heavy bombardment mission over enemy-occupied continental Europe. Coming off the target he was attacked by approximately 20 enemy fighters, shot out of formation, and his plane severely crippled. Eight crewmembers were wounded; the copilot was killed by a 20-mm shell. One engine was on fire, the controls shot away, and 1st Lt. Lawley seriously and painfully wounded about the face. Forcing the copilot's body off the controls, he brought the plane out of a steep dive, flying with his left hand only. Blood covered the instruments and windshield and visibility was impossible. With a full bomb load the plane was difficult to maneuver and the bombs could not be released because the racks were frozen. After the order to bail out had been given, one of the waist gunners informed the pilot that two crewmembers were so severely wounded that it would be impossible for them to bail out. With the fire in the engine spreading, the danger of an explosion was imminent. Because of the helpless condition of his wounded crewmembers 1st Lt. Lawley elected to remain with the ship and bring them to safety if it was humanly possible, giving the other crewmembers the option of bailing out. Enemy fighters again attacked but by using masterful evasive action he managed to lose them. One engine again caught on fire and was extinguished by skillful flying. First Lt. Lawley remained at his post, refusing first aid until he collapsed from sheer exhaustion caused by loss of blood, shock, and the energy he had expended in keeping control of his plane. He was revived by the bombardier and again took over the controls. Coming over the English coast one engine ran out of gasoline and had to be feathered. Another engine started to burn and continued to do so until a successful crash landing was made on a small fighter base. Through his heroism and exceptional flying skill 1st Lt. Lawley rendered outstanding distinguished and valorous service to our nation.

Epilogue

Lawley joined the Army Air Forces from Birmingham, Alabama in April 1942, and by February 20, 1944, was a first lieutenant serving as a pilot in the 364th Bomb Squadron, 305th Bomb Group. On that day, during a bombing mission over Nazi-controlled Europe, his B-17 Flying Fortress came under attack by enemy fighter aircraft. With his plane severely damaged and on fire, he ordered his crew to parachute to safety. Finding that two crewmen were too badly injured to bail out, Lawley chose to remain in the aircraft and, despite his own serious wounds, attempt to pilot it into friendly territory. Although weakened from loss of blood and shock, he was able to make a successful crash landing in England. Lawley recovered from his wounds and, on August 8, 1944, was awarded the Medal of Honor.

Lawley reached the rank of colonel before leaving the Air Force in 1972. He died at age 78 and was buried in Greenwood Cemetery, Montgomery, Alabama.

JAKE W. LINDSEY SR.

Rank and Organization: Technical Sergeant (Highest rank: 1st Lieutenant); Company C, 16th Infantry, 1st Infantry Division; **Military Service Branch**: U. S. Army; **Conflict**: World War II; **Medal of Honor Action Date**: November 16, 1944; **Medal of Honor Action Place**: Near Hamich, Germany

Presentation: May 21, 1945 at a Joint Session of Congress, Washington, DC, presented by President Harry S. Truman

Accredited to: Lucedale, George County, Mississippi

Born: May 1, 1921, Isney, Choctaw County, Alabama

Died: July 18, 1988, Waynesboro, Mississippi

Buried: White House Cemetery (MH), Waynesboro, Mississippi

Citation

For gallantry and intrepidity in action at the risk of his life above and beyond the call of duty. Technical Sergeant Jake W. Lindsey led a platoon, reduced to six of its original strength of forty, in the attack on an enemy position near Hamich, Germany, 16 November 1944. His men had captured their objective and were digging in when counterattacked by a German infantry company and five tanks. Armed with a rifle and grenades, TSgt. Lindsey took position on the left and in advance of the remnant of his platoon, and though exposed to heavy rifle, machine-gun and tank fire, beat off repeated enemy attacks. Tanks moved to within 50 yards of him but were forced to withdraw because of his accurate rifle and grenade fire. After driving off the tanks, he knocked out two machine guns to his front. Though painfully wounded, TSgt. Lindsey continued firing and throwing grenades until his ammunition was expended. An enemy squad attempted to set up a machine gun 50 yards from him. Unmindful of his wounds and enemy fire, he rushed these eight German soldiers, singlehandedly closed with them, killed three with his bayonet, and captured three, the two others escaping. In his fearlessness, inspiring courage and superb leadership, TSgt. Lindsey carried on a brilliant defense of his platoon's hard-won ground, securing the position and inflicting heavy casualties on the numerically superior enemy.

Epilogue

Lindsey joined the Army from Huntington, West Virginia in February 1940 after working as a plasterer as a civilian, and by November 16, 1944 was serving as a technical sergeant in the 16th Infantry Regiment, 1st Infantry Division. On that day, near Hamich, Germany, he held a position in front of his platoon during an enemy counterattack and, although wounded, engaged a group of Germans in hand to hand combat. For these actions, he was awarded the Medal of Honor six months later.

Lindsey is unique among Medal of Honor recipients in that he personally received his medal before a joint session of Congress. President Harry Truman and General of the Army George C. Marshall presented Lindsey with his award on May 21, 1945. Footage of this ceremony is viewable on YouTube.

Lindsey later served in the Korean War and left the Army while a second lieutenant. He died at age 67 and was buried in Whitehouse Cemetery, Clara, Mississippi.

JACK L TREADWELL

Rank and Organization: Captain (Rank at the time of action: 1st Lieutenant), Company F, 180th Infantry, 54th Infantry Division; **Military Service Branch**: U. S. Army; **Conflict**: World War II; **Medal of Honor Action Date**: March 18, 1945; **Medal of Honor Action Place**: Near Nieder-Wurzbach, Germany

Presentation: August 23, 1945 at The White House, presented by President Harry S. Truman

Accredited to: Snyder, Kiowa County, Oklahoma

Born: March 31, 1919, Ashland, Clay County, Alabama

Died: December 12, 1977

Buried: Fort Sill Cemetery (MH), Fort Sill, Oklahoma

Citation

Capt. Treadwell (then 1st Lt.), commanding officer of Company F, near Nieder-Wurzbach, Germany, in the Siegfried line, singlehandedly captured six pillboxes and 18 prisoners. Murderous enemy automatic and rifle fire with intermittent artillery bombardments had pinned down his company for hours at the base of a hill defended by concrete fortifications and interlocking trenches. Eight men sent to attack a single point had all become casualties on the bare slope when Capt. Treadwell, armed with a submachine gun and hand grenades, went forward alone to clear the way for his stalled company. Over the terrain devoid of cover and swept by bullets, he fearlessly advanced, firing at the aperture of the nearest pillbox and, when within range, hurling grenades at it. He reached the pillbox, thrust the muzzle of his gun through the port, and drove four Germans out with their hands in the air. A fifth was found dead inside. Waving these prisoners back to the American lines, he continued under terrible, concentrated fire to the next pillbox and took it in the same manner. In this fort he captured the commander of the hill defenses, whom he sent to the rear with the other prisoners. Never slackening his attack, he then ran across the crest of the hill to a third pillbox, traversing this distance in full view of hostile machine gunners and snipers. He was again successful in taking the enemy position. The Germans quickly fell prey to his further rushes on three more pillboxes in the confusion and havoc caused by his whirlwind assaults and capture of their commander. Inspired by the electrifying performance of their leader, the men of Company F stormed after him and overwhelmed resistance on the entire hill, driving a wedge into the Siegfried line and making it possible for their battalion to take its objective. By his courageous willingness to face nearly impossible odds and by his overwhelming one-man offensive, Capt. Treadwell reduced a heavily fortified, seemingly impregnable enemy sector.

Epilogue

Treadwell was born on March 30, 1919, in Ashland, Alabama. He graduated from Snyder High School in 1937 and attended Southwestern State College in Weatherford, Oklahoma in 1937 and 1938. Treadwell enlisted in the U.S. Army from Snyder, Oklahoma on January 29, 1941, and was trained as an infantryman.

He served with the 180th Infantry Regiment of the 45th Infantry Division throughout World War II. He deployed with the unit to North Africa in June 1943, participating in the amphibious assault on Sicily in July 1943, Salerno in September 1943, Southern France in August 1944 and also served in Alsace and Rhineland Campaigns. He received a battlefield commission on March 23, 1944.

By March 18, 1945 was serving as a first lieutenant in command of Company F, 180th Infantry Regiment, 45th Infantry Division. During a battle on that day, near Nieder-Wurzbach in Germany, Treadwell single-handedly captured six German bunkers. He was subsequently promoted to captain and, on September 14, 1945, awarded the Medal of Honor. Treadwell was wounded in March 1945, and after hospitalization, he returned to duty with the Infantry School at Fort Benning in March 1946.

Treadwell's next assignment was with the 7th Infantry Regiment at Fort Devens, from April 1948 to September 1949, followed by attending the Armored School at Fort Knox, from September 1949 to August 1950. Treadwell served as a company commander and executive officer with the 350th Infantry Regiment in Austria from August 1950 to December 1952, and then as Aide-de-Camp to the commander of First United States Army at Fort Jay from December 1952 to July 1954.

He next completed Command and General Staff College at Fort Leavenworth and then served as the headquarters commandant of the Army Forces on the Eniwetok Atoll in the Marshall Islands from September 1955 to September 1956. Treadwell returned to the United States Army Infantry School at Fort Benning from September 1956 to August 1958, followed by Armed Forces Staff College at Norfolk, Virginia, from August 1958 to February 1959.

His next assignment was at Fort Sam Houston, from February 1959 to August 1961, followed by Army War College at from August 1961 to September 1962. Treadwell next completed a degree program at the University of Omaha from September 1962 to July 1963, and then served on as the staff of United States Army Europe and as a battalion commander in West Germany from July 1963 to July 1966. He served as the Chief of the Army Infantry Center at Fort Benning from July to November 1966, and then served as commander of the 197th Infantry Brigade at Fort Benning until August 1968.

Treadwell served as Chief of Staff of the Americal Division in Vietnam from October 1968 to March 1969, and then as commander of the 11th Infantry Brigade in Vietnam from March to September 1969. He served in Vietnam where he made over 100 parachute jumps. He participated in three campaigns in Vietnam: the Vietnam Counteroffensive Phase 5, Vietnam Counteroffensive Phase 6 and the Tet Offensive.

In June 1970, he was presented the Oklahoma Distinguished Service Cross in recognition of extraordinary heroism involving great personal danger and risk of life in the line of military duty and for outstanding service to the Oklahoma Army National Guard during the period 1941 till 1970. He was presented the state's highest decoration for extraordinary heroism was made by then Governor Dewey Bartlett.

After Vietnam, Treadwell served with Headquarters Fourth United States Army (redesignated Fifth United States Army in June 1971) at Fort Sam Houston from October 1969 to June 1972, followed by service as Senior Army Advisor to the U.S. Army Reserve at Columbia, South Carolina, from July 1972.

After more than 33 years service, Treadwell retired on Wednesday, Feb. 27, 1974 in retirement ceremonies held at the Fort Jackson Officer's Open Mess at 10:30 A.M. Major General William H. Blakefield, commander, Readiness Region II, presented Treadwell with the third oak leaf cluster for his Legion of Merit, for exceptional meritorious conduct from July 1972 through February 1974, while assigned as the Senior Army Advisor, 120th Army Reserve Command, U.S. Army Readiness Region III. Treadwell was also presented a Certificate of Appreciation from Creighton Abrams, the Chief of Staff of the United States Army.

Treadwell was married to Maxine Johnson of Mooresville, Indiana, an Army nurse whom he met while he was recuperating in France from his most serious wound (a ricochet through his neck and chest).

After retiring, Treadwell and his wife settled in Oklahoma, where they planned to raise horses. The Treadwells have three daughters, two of whom are married to officers in the U.S. Army. Treadwell died aged 58 on 12 December 1977 following open-heart bypass surgery. He is buried with full military honors at the Fort Sill Post Cemetery near Lawton, Oklahoma.

CHARLES L. THOMAS

<u>POSTHUMOUS AWARD</u>

Rank and Organization: Lieutenant, Company C, 614 Tank Destroyer Battalion; **Military Service Branch**: U. S. Army; **Conflict**: World War II; **Medal of Honor Action Date**: December 14, 1944; **Medal of Honor Action Place**: Climbach, France

Presentation: January 13, 1997 at The White House, presented by President William J. Clinton

Accredited to: Detroit, Cook County, Michigan

Born: April 17, 1920, Birmingham, Jefferson County, Alabama

Died: February 15, 1980, Wayne, Michigan

Buried: Fort Sill Cemetery (MH), Fort Sill, Oklahoma

Citation

For conspicuous gallantry and intrepidity at the risk of his life above and beyond the call of duty: Then Lieutenant Charles L. Thomas distinguished himself by extraordinary heroism in action on 14 December 1944. One platoon of Company C, 614th Tank Destroyer Battalion, was designated as the lead element in a task force formed to storm and capture the village of Climbach, France. Lieutenant Thomas, the Commanding Officer of Company C, realized, with the obscurity of information regarding the enemy and a complete lack of reconnaissance, the mission would be an extremely dangerous one. Fully cognizant of the danger, Lieutenant Thomas volunteered to command the selected platoon of his company and ride in the column's leading vehicle - a highly maneuverable, but equally vulnerable, M-20 scout car. Lieutenant Thomas knew that if there was a concentration of enemy armor in the village, as was believed, he would absorb the initial shock of the first enemy resistance. The task force left Preuschdorf, France, at 1023 hours, and proceeded to advance in column toward Climbach. Lieutenant Thomas in his scout car stayed well in front of the column. At 1400 hours, upon reaching the high ground southeast of the village, Lieutenant Thomas experienced initial contact with the enemy. As his scout car advanced to an exposed position on the heights, he received intense direct fire from enemy artillery, self-propelled guns, and small arms at a range of seven hundred yards. The first burst of hostile fire disabled the scout car and severely wounded Lieutenant Thomas. He immediately signaled the column to halt. Before leaving the wrecked vehicle, Lieutenant Thomas and the crew found themselves subjected to a veritable hail of enemy fire. Lieutenant Thomas received multiple gunshot wounds in his chest, legs, and left arm. In spite of the intense pain caused by his wounds, Lieutenant Thomas ordered and directed the dispersion and emplacement of his first two antitank guns. In a few minutes these guns were effectively returning the enemy fire. Realizing that it would be impossible for him to remain in command of the platoon because of his injuries, Lieutenant Thomas then signaled for the platoon commander to join him. Lieutenant Thomas then thoroughly oriented him as to the enemy gun positions, his ammunition status, and the general situation. Although fully cognizant of the probable drastic consequences of not receiving prompt medical attention, Lieutenant Thomas refused evacuation until he felt certain that his junior officer was in full control of the situation. Only then did Lieutenant Thomas allow his evacuation to the rear. Throughout the action, Lieutenant Thomas displayed magnificent personal courage and a complete disregard for his own safety. His extraordinary heroism spurred the soldiers of the platoon to a fierce determination to triumph, and resulted in a mass display of heroism by them. Lieutenant Thomas' intrepid actions throughout the operation reflect the highest traditions of military service.

Epilogue

Thomas was born in Birmingham, Alabama. He grew up in Detroit and graduated from Cass Technical High School in 1938. Thomas then worked as a molder for the Ford Motor Company (Ford River Rogue factory) with his father, and was a student at Wayne State University studying mechanical engineering.

He was drafted, and entered into the U.S. Army on January 20, 1942. He went to basic training at Camp Wolters in Texas, and afterwards was assigned to the infantry. Next, he was transferred to the Officer Candidate School at Camp Carson in Colorado where black soldiers were sent to be officers in the 614th Tank Destroyer Battalion. He was commissioned as a Second Lieutenant on March 11, 1943, and was assigned to be the commander of Company C, 614th Tank Destroyer Battalion at Fort Hood in Killeen, Texas. On August 27, 1944, he deployed with the 614th to England, arriving on September 7. On October 8, the 614th was on Utah Beach in Normandy, France. The 614th led by Lieutenant Colonel Frank S. Pritchard, then would join General Patton's Third Army in Metz, France. The 614th saw its first combat on November 28. On December 5, the 614th was attached to 411th Infantry Regiment, and on December 6, the 614th was attached to the 103rd Infantry Division.

On December 14, 1944, 1st Lt. Thomas volunteered to lead 3rd Platoon, C Company, 614th Tank Destroyer Battalion in a task force named "Task Force Blackshear" to storm and capture the village of Climbach, a strategically important town which was five miles from the German border. The task force spearheaded by Thomas' M20 scout car (modified M8), consisted of a platoon of Sherman tanks from the 47th Armored Battalion, 14th Armored Division, a platoon of F Company (riding on tanks), 411th Infantry, 103rd Infantry Division, 3rd Platoon, C Company, 614th Tank Destroyer Battalion, rest of F Company, 411th Infantry, and a heavy weapons platoon. Approaching Climbach was uphill, Thomas' armored scout car was knocked out by enemy fire from the German 21st Panzer Division, and he was wounded.

Thomas helped his crew out of the vehicle, but as he left the car's protection, he was again wounded in the chest, legs and arms. Despite his wounds, Thomas directed the dispersal and emplacement of the anti-tank guns, which then returned fire and covered the attempt by the rest of the task force to outflank the defenders. He briefed the 3rd Platoon leader of C Company, a first lieutenant, on the general situation, and only when he was sure the situation was under control did he allow himself to be evacuated. 3rd Platoon, C Company continued to fight for four hours, losing two of its four guns and over half its men as casualties (3 dead, 17 wounded).

Captain Thomas receiving the Distinguished Service Cross (revoked 1997 for MOH) in 1945

The "valorous conduct" of the platoon, "in the face of overwhelming odds enabled the task force to capture its objective", the village of Climbach, and forced the defenders to withdraw to the Siegfried Line. 3rd Platoon, C Company, 614th Tank Destroyer Battalion, was awarded a Distinguished Unit Citation, the first black combat unit, and the first unit attached to the 103rd Division to be so honored. Its soldiers received four Silver Stars and nine Bronze Stars. Captain Thomas received the Distinguished Service Cross on February 20, 1945, and returned home a hero, though he played down his role – "I know I was sent out to locate and draw the enemy fire, but I didn't mean to draw that much." Thomas remained in the Army, and retired with the rank of major on August 10, 1947.

Thomas married in 1949, and his wife and he had two children. He went to work as a missile technician at Selfridge Air Force Base and later as a computer programmer for the Internal Revenue Service. He died of cancer on February 15, 1980. He was buried in Westlawn Cemetery in Wayne, Michigan.

In the early 1990s, it was determined that Black soldiers had been denied consideration for the Medal of Honor (MOH) in World War II because of their race. In 1993, the U.S. Army had contracted Shaw University in Raleigh, North Carolina, to research and determine if there was racial disparity in the review process for recipients of the MOH. The study commissioned by the U.S. Army, described systematic racial discrimination in the criteria for awarding decorations during World War II. After an exhaustive review of files, the study recommended in 1996 that ten Black Americans who served in World War II be awarded the MOH. In October of that year, Congress passed legislation that would allow President Clinton to award the Medal of Honor to these former soldiers. Seven of the ten including Thomas were approved, and awarded the MOH (six had Distinguished Service Crosses revoked and upgraded to the MOH) on January 12, 1997. On January 13, 1997, President Clinton presented the MOH to the seven Black Americans; Major Thomas and five others were posthumously presented the MOH. A niece of Thomas accepted his MOH during the ceremony. Vernon Baker was the only living recipient of the Medal at the time.

DAVID McCAMPBELL

Rank and Organization: Commander, Air Group 15, U.S.S. Essex; **Military Service Branch**: U. S Navy; **Conflict**: World War II; **Medal of Honor Action Date**: October 24-25, 1944; **Medal of Honor Action Place**: Battle of the Philippine Sea & Battle of Leyte Gulf, Philippine Island

Presentation: January 10, 1945, The White House, presented by President Harry S. Truman

Accredited to: West Palm Beach, Palm Beach County, Florida

Born: January 16, 1910, Bessemer, Jefferson County, Alabama

Died: June 30, 1996, West Palm Beach, Florida

Buried: Arlington National Cemetery (MH) (60-3150), Arlington, Virginia

Citation

For conspicuous gallantry and intrepidity at the risk of his life above and beyond the call of duty as commander, Air Group 15, during combat against enemy Japanese aerial forces in the first and second battles of the Philippine Sea. An inspiring leader, fighting boldly in the face of terrific odds, Comdr. McCampbell led his fighter planes against a force of 80 Japanese carrier-based aircraft bearing down on our fleet on 19 June 1944. Striking fiercely in valiant defense of our surface force, he personally destroyed seven hostile planes during this single engagement in which the outnumbering attack force was utterly routed and virtually annihilated. During a major fleet engagement with the enemy on 24 October, Comdr. McCampbell, assisted by but one plane, intercepted and daringly attacked a formation of 60 hostile land-based craft approaching our forces. Fighting desperately but with superb skill against such overwhelming airpower, he shot down nine Japanese planes and, completely disorganizing the enemy group, forced the remainder to abandon the attack before a single aircraft could reach the fleet. His great personal valor and indomitable spirit of aggression under extremely perilous combat conditions reflect the highest credit upon Comdr. McCampbell and the U.S. Naval Service.

Epilogue

McCampbell was born in Bessemer, Alabama, and raised in West Palm Beach, Florida. He attended the Staunton Military Academy, in Virginia, and one year at the Georgia School of Technology, in Atlanta, before his appointment to the United States Naval Academy in 1929, where he graduated with the class of 1933 with a degree in marine engineering.

McCampbell's naval career actually began with a dismissal. Graduating from the U.S. Naval Academy in depression-era 1933, he was rewarded with an honorable discharge from a navy without funds. But in June 1934 McCampbell was called back and commissioned as an ensign in the U.S. Naval Reserve on June 1, 1934. He went on active duty on June 14, 1934, and served aboard the heavy cruiser USS *Portland* from June 1934 to June 1937 before he started flight training at Naval Air Station Pensacola, Florida. He received his "wings of gold" on April 21, 1938 and was assigned to Fighting Squadron Four (VF-4) on the aircraft carrier USS *Ranger* to May 1940.

McCampbell served as a landing signal officer (LSO) from May 1940, surviving the sinking of the carrier USS *Wasp* (CV-7) by a Japanese submarine near Guadalcanal on September 15, 1942. He returned to the United States, was promoted to Lieutenant Commander, and was stationed at Naval Air Station Melbourne, Florida as LSO Instructor until August 1943.

McCampbell formed Fighter Squadron 15 (VF-15) on September 1, 1943 and led the squadron before being reassigned as Commander of Air Group 15 (CAG-15) in February 1944 to September 1944. As Commander, Carrier Air Group (CAG) 15, he was Commander of the *Essex* Air Group (fighters, bombers, and torpedo bombers) when the group was aboard the aircraft carrier USS *Essex*. From April to November 1944, his group saw six months of combat and participated in two major air-sea battles, the First and Second Battles of the Philippine Sea. During the more than 20,000 hours of air combat operations before it returned to the United States for a rest period, Air Group 15 destroyed more enemy planes (315 airborne and 348 on the ground) and sank more enemy shipping than any other Air Group in the Pacific War. Air Group 15's attacks on the Japanese in the Marianas and at Iwo Jima, Taiwan, and Okinawa were key to the success of the "island hopping" campaign.

Grumman F6F-5 Hellcat displayed at the National Naval Aviation Museum, painted as a replica of McCampbell's *Minsi III* to include Japanese flags representing 34 aerial victories.

In addition to his duties as commander of the "Fabled Fifteen", then Commander McCampbell became the Navy's "ace of aces" during the missions he flew in 1944. McCampbell entered combat on May 14 and flew at least four Grumman F6F Hellcats while aboard the *Essex*: an F6F-3 named *Monsoon Maiden* (damaged by AA, removed from service on 20 May 1944), an F6F-3 named *The Minsi* (10+1⁄2 kills), an F6F-5 named *Minsi II*, and an F6F-5 named *Minsi III* (Bureau Number 70143), in which he scored the last 23+1⁄2 of his 34 kills.

On June 19, 1944, during the "Marianas Turkey Shoot," Commander McCampbell shot down five Japanese Yokosuka D4Y 'Judy' dive-bombers, to become an "ace in a day". Later that afternoon, during a second sortie, McCampbell downed two Mitsubishi A6M 'Zekes' over Guam.

On October 24, 1944, in the initial phase of the Battle of Leyte Gulf, in the Philippines, he became the only American airman to achieve "ace in a day" status twice. McCampbell and his wingman attacked a Japanese force of 60 aircraft. McCampbell shot down nine—seven Zeros and two Oscars—setting a U.S. single-mission aerial combat record. During this same action, his wingman downed another six Japanese warplanes. When he landed his Grumman F6F Hellcat aboard USS *Langley* (the flight deck of *Essex* wasn't clear), his six machine guns had just two rounds remaining, and his airplane had to be manually released from the arrestor wire due to complete fuel exhaustion. Commander McCampbell received the Medal of Honor for both actions, becoming the only Fast Carrier Task Force pilot to be so honored.

HENRY EUGENE “RED” ERWIN

Rank and Organization: Staff Sergeant, 52 Bombardment Squadron, 29th Bombardment Group, 314th Bombardment Wing, 20th Air Force; **Military Service Branch**: U. S. Army; **Conflict**: World War II; **Medal of Honor Action Date**: April 12, 1945; **Medal of Honor Action Place**: Koriyama, Japan

Presentation: April 19, 1945, Fleet Hospital103, Guam, presented by Gen. Curtis Lemay

Accredited to: Bessemer, Jefferson County, Alabama

Born: May 8, 1921, Adamsville, Jefferson County, Alabama

Died: January 16, 2002, Leeds, Alabama

Buried: Elmwood Cemetery (MH) (2-E1/2-279)

Citation

He was the radio operator of a B-29 airplane leading a group formation to attack Koriyama, Japan. He was charged with the additional duty of dropping phosphorus smoke bombs to aid in assembling the group when the launching point was reached. Upon entering the assembly area, aircraft fire and enemy fighter opposition was encountered. Among the phosphorus bombs launched by S/Sgt. Erwin, one proved faulty, exploding in the launching chute, and shot back into the interior of the aircraft, striking him in the face. The burning phosphorus obliterated his nose and completely blinded him. Smoke filled the plane, obscuring the vision of the pilot. S/Sgt. Erwin realized that the aircraft and crew would be lost if the burning bomb remained in the plane. Without regard for his own safety, he picked it up and feeling his way, instinctively, crawled around the gun turret and headed for the copilot's window. He found the navigator's table obstructing his passage. Grasping the burning bomb between his forearm and body, he unleashed the spring lock and raised the table. Struggling through the narrow passage he stumbled forward into the smoke-filled pilot's compartment. Groping with his burning hands, he located the window and threw the bomb out. Completely aflame, he fell back upon the floor. The smoke cleared, the pilot, at 300 feet, pulled the plane out of its dive. S/Sgt. Erwin's gallantry and heroism above and beyond the call of duty saved the lives of his comrades.

Epilogue

Erwin was born on May 8, 1921, in Adamsville, Alabama. Like many of his generation he grew up in poverty and lost his father at an early age. However he had very strong religious faith which he discussed on a History Channel documentary on Medal of Honor recipients in 1999. Erwin said "I called on the Lord to help me and He has never let me down". Erwin joined the Army Reserve from nearby Bessemer on July 27, 1942. Called to active duty as an aviation cadet in the Army Air Forces on February 3, 1943, he trained as a pilot in Ocala, Florida, but washed out due to "flying deficiency". He was instead transferred to technical school at Keesler Air Force Base, Mississippi, as a private first class in July of that year. He completed further radio operator and radio mechanic training in Sioux Falls, South Dakota, and Madison, Wisconsin, until his graduation in 1944.

Assigned to the 52d Bombardment Squadron, 29th Bombardment Group, Twentieth Air Force, in Dalhart, Texas, Erwin and his unit left for the Asia-Pacific theater in early 1945. From February 25 to April 1 of that year, they participated in a series of un-escorted bombing strikes against cities in the heart of Japan. For these missions, Erwin, by then a staff sergeant, received two Air Medals. Erwin served as a Radio Operator Gunner during his time in the service. He was Medically Discharged from the United States Army Air Force on October 27, 1947 having served 5 years, 2 months, and 12 days. During World War II, he participated in the Air Raids on Japan Campaign and the Western Pacific Campaign. For his Military Service during World War II, he was awarded the Army Air Force Aircrew Member Badge, the Medal of Honor, American Campaign Medal, Asiatic-Pacific Campaign Medal with 2 3/16" bronze stars, Army Good Conduct Medal, Purple Heart Medal, Air Medal with 1 bronze oak leaf cluster, and the World War II Victory Medal.

On April 12, 1945, Erwin, called "Red" by his crewmates, was serving as the radio operator aboard a Boeing B-29 Superfortress named *City of Los Angeles,* piloted by Captain George Simeral. The plane was in formation for a low-level attack on a chemical plant at Koriyama, 120 miles (190 km) north of Tokyo, on their 11th combat mission. Along with their primary jobs, the twelve B-29 crew members had additional duties to perform. Erwin's was to drop phosphorus smoke bombs through a chute in the aircraft's floor when the lead plane reached a designated assembly area. He was given the signal to drop the bombs when the aircraft was just off the south coast of Japan and under attack by anti-aircraft fire and Japanese fighters.

Erwin pulled the pin and released a bomb into the chute, but the fuse malfunctioned and ignited the phosphorus prematurely, burning at

1,100 degrees. The canister flew back up the chute and into Erwin's face, blinding him, searing off one ear and obliterating his nose. Phosphorus pentoxide smoke immediately filled the aircraft, making it impossible for the pilot to see his instrument panel. Erwin was afraid the bomb would burn through the metal floor into the bomb bay. Completely blind, he picked it up and feeling his way, crawled around the gun turret and headed for the copilot's window. His face and arms were covered with ignited phosphorus and his path was blocked by the navigator's folding table, hinged to the wall but down and locked. The navigator had left his table to make a sighting. Erwin couldn't release the table's latches with one hand, so he grabbed the white-hot bomb between his bare right arm and his ribcage. In the few seconds it took to raise the table, the phosphorus burned through his flesh to the bone. His body on fire, he stumbled into the cockpit, threw the bomb out the window and collapsed between the pilot's seats.

The smoke cleared enough for Simeral to pull the B-29 out of a dive at 300 feet (91 m) above the water and turn toward Iwo Jima, where Erwin could be given emergency treatment. His crew members extinguished his burning clothes and administered first aid, but whenever Erwin's burns were uncovered, phosphorus embedded in his skin would begin to smolder; white phosphorus is known to cause extremely terrible wounds, as the burning chemical cannot be extinguished if oxygen is present, and will continue to burn through flesh until it consumes itself or is extracted. It is also toxic. Although in excruciating pain, he remained conscious throughout the flight and spoke only to inquire about the safety of the crew. Once on Iwo Jima the medical personnel who examined Erwin expected him to die.

Army Air Force officials, led by Major General Curtis LeMay and Brigadier General Lauris Norstad, approved Erwin's award of the Medal of Honor in a matter of hours, so a presentation could be made while he still lived. A medal was flown from Hawaii to Guam and presented to him in the hospital there.

However, Erwin survived his burns. He was flown back to the United States, and after 30 months and 41 surgeries, his eyesight was restored and he regained use of one arm. He received a disability discharge as a master sergeant in October 1947. In addition to the Medal of Honor and two Air Medals received earlier in 1945, he was also awarded the Purple Heart, the World War II Victory Medal, the

American Campaign Medal, three Good Conduct Medals, the Asiatic-Pacific Campaign Medal with two bronze campaign stars (for participation in the Air Offensive Japan and Western Pacific campaigns), and the Distinguished Unit Citation Emblem.

Erwin in 1995 participating in the ceremony commemorating the 50th anniversary of the end of World War II

For 37 years, Erwin served as a benefits counselor at the veterans' hospital in Birmingham, Alabama. In 1951, his story was included in the movie *The Wild Blue Yonder;* Erwin was portrayed by Dave Sharpe.

In 1997, the Air Force created the Henry E. Erwin Outstanding Enlisted Aircrew Member of the Year Award. It is presented annually to an airman, noncommissioned officer and senior noncommissioned officer in the flight engineering, loadmaster, air surveillance and related career fields. It is only the second Air Force award named for an enlisted person.

Erwin died at his home on January 16, 2002, and was buried at Elmwood Cemetery in Birmingham, Alabama. His son, Hank Erwin, became an Alabama state senator.

Korean War Recipients of the Medal of Honor

HAROLD E WILSON

Rank and Organization: Technical Sergeant 9Highest Rank: Chief Warrant Officer 4); #rd Platoon, Company G, 3rd Battalion, 1st Marines, 1st Marine Division; **Military Service Branch**: U. S. Marine Corps; **Conflict**: Korean War; **Medal of Honor Action Date**: April 23, 1951; **Medal of Honor Action Place**: Hill 902 near Hwachon Reservoir, Korea

Presentation: April 11, 1952, The White House, presented by President Harry S. Truman

Accredited to: Birmingham, Jefferson County, Alabama

Born: December 5, 1921, Birmingham, Jefferson County, Alabama

Died: March 29, 1998, West Columbia, South Carolina

Buried: Woodridge Memorial Park Cemetery (MH) (2-191-1), Lexington, South Carolina

Citation

For gallantry and intrepidity at the risk of his life above and beyond the call of duty while serving as platoon sergeant of a rifle platoon attached to Company G, in action against enemy aggressor forces on the night from 23 to 24 April 1951. When the company outpost was overrun by the enemy while his platoon, firing from hastily constructed foxholes, was engaged in resisting the brunt of a fierce mortar, machine-gun, grenade, and small-arms attack launched by hostile forces from high ground under cover of darkness, T/Sgt. Wilson braved intense fire to assist the survivors back into the line and to direct the treatment of casualties. Although twice wounded by gunfire, in the right arm and the left leg, he refused medical aid for himself and continued to move about among his men, shouting words of encouragement. After receiving further wounds in the head and shoulder as the attack increased in intensity, he again insisted upon remaining with his unit. Unable to use either arm to fire, and with mounting casualties among our forces, he resupplied his men with rifles and ammunition taken from the wounded. Personally reporting to his company commander on several occasions, he requested and received additional assistance when the enemy attack became even more fierce and, after placing the reinforcements in strategic positions in the line, directed effective fire until blown off his feet by the bursting of a hostile mortar round in his face. Dazed and suffering from concussion, he still refused medical aid and, despite weakness from loss of blood, moved from foxhole to foxhole, directing fire, resupplying ammunition, rendering first aid, and encouraging his men. By his heroic actions in the face of almost certain death, when the unit's ability to hold the disadvantageous position was doubtful, he instilled confidence in his troops, inspiring them to rally repeatedly and turn back the furious assaults. At dawn, after the final attack had been repulsed, he personally accounted for each man in his platoon before walking unassisted one-half mile to the aid station where he submitted to treatment. His outstanding courage, initiative, and skilled leadership in the face of overwhelming odds were contributing factors in the success of his company's mission and reflect the highest credit upon T/Sgt. Wilson and the U.S. Naval Service.

Epilogue

Wilson was born on December 5, 1921, in Birmingham, Alabama, and attended Central Park Elementary and Ensley High School. He enlisted in the Marine Corps Reserve and was assigned to active duty on April 20, 1942. During World War II, he served overseas for 27 months and was stationed on Midway Island. In addition to his Pacific service, he was stationed at Parris Island, South Carolina; Camp Lejeune, North Carolina; and Portsmouth, Virginia. He was honorably discharged on October 20, 1945, with the rank of sergeant.

Two years later, he re-entered the Organized Marine Corps Reserve, joining the Birmingham unit. Recalled to active duty in August 1950 after the outbreak of the Korean War, he was assigned to Company G, 3rd Battalion, 1st Marines, 1st Marine Division. Arriving with his unit in Korea in November 1950, he participated in the Wonsan landing, and later fought in the Battle of Chosin Reservoir and in the first United Nations counteroffensive. Following the April 23–24, 1951, action in which he was wounded and earned the Medal of Honor, he was evacuated to Yokosuka Naval Hospital in Japan and five months later returned to the United States. He was awarded a meritorious promotion to master sergeant in June 1951 and commissioned a warrant officer in August 1952. In December 1962, he assumed the post of adjutant at the Marine Corps Engineer Schools, Camp Lejeune, and the following December was assigned to Force Troops, Fleet Marine Force, Atlantic, serving as adjutant and personnel officer of the 2nd Tank Battalion.

Wilson was also a veteran of the Vietnam War, serving with Marine Aircraft Group 13 prior to being assigned as 6th Marine Corps District Personnel Officer in November 1968. He retired from the Marine Corps in February 1972.

Wilson died in Lexington, South Carolina on March 29, 1998. He is buried in Woodridge Memorial Park, Lexington, South Carolina.

ALFORD L. MCLAUGHLIN

Rank and Organization: Private First Class (Highest Rank: Master Sergeant), Company L, 3rd Battalion, 5th Marines, 1st Marine Division; **Military Service Branch**: U. S. Marine Corps; **Conflict**: Korean War; **Medal of Honor Action Date**: September 4, 1952; **Medal of Honor Action Place**: Outpost Bruce, Korea

Presentation: October 27, 1953, The White House, presented by President Dwight D. Eisenhower

Accredited to: Leeds, Jefferson County, Alabama

Born: March 18, 1928, Leeds, Jefferson County, Alabama

Died: January 15, 1977, Leeds, Alabama

Buried: Mount Hebron Cemetery (MH), Leeds, Alabama

Citation

For conspicuous gallantry and intrepidity at the risk of his life above and beyond the call of duty while serving as a machine gunner of Company L, in action against enemy aggressor forces on the night of 4-5 September 1952. Volunteering for his second continuous tour of duty on a strategic combat outpost far in advance of the main line of resistance, Pfc. McLaughlin, although operating under a barrage of enemy artillery and mortar fire, set up plans for the defense of his position which proved decisive in the successful defense of the outpost. When hostile forces attacked in battalion strength during the night, he maintained a constant flow of devastating fire upon the enemy, alternately employing two machine guns, a carbine, and hand grenades. Although painfully wounded, he bravely fired the machine guns from the hip until his hands became blistered by the extreme heat of the weapons and, placing the guns on the ground to allow them to cool, continued to defend the position with his carbine and grenades. Standing up in full view, he shouted words of encouragement to his comrades above the din of battle and, throughout a series of fanatical enemy attacks, sprayed the surrounding area with deadly fire, accounting for an estimated 150 enemy dead and 50 wounded. By his indomitable courage, superb leadership, and valiant fighting spirit in the face of overwhelming odds, Pfc. McLaughlin served to inspire his fellow marines in their gallant stand against the enemy and was directly instrumental in preventing the vital outpost from falling into the hands of a determined and numerically superior hostile force. His outstanding heroism and unwavering devotion to duty reflect the highest credit upon himself and enhance the finest traditions of the U.S. Naval Service.

Epilogue

Born on March 18, 1928, in Leeds, Alabama, McLaughlin attended school in Leeds until 1944 and then enlisted in the Marine Corps on May 3, 1945. After completing recruit training at Parris Island South Carolina, he served at Camp Lejeune, North Carolina, until embarking for Guam in November 1945. From Guam he was ordered to Japan in March 1946, serving in the occupation of that country until August 1947. He then served with the 4th Marines, participating in Caribbean maneuvers from January to March 1948, and again from February to March 1949. He served in the Mediterranean from September 1948 to January 1949.

Private First Class McLaughlin was assigned to the Marine Detachment at the U.S. Naval Disciplinary Barracks, Portsmouth, New Hampshire, from June 1949 until September 1951. He was next assigned to Camp Pendleton, California, for further training before leaving for Korea in February 1952. He fought in the second Korean winter, the summer-fall defense of 1952, and in the third Korean winter before he left Korea in January 1953.

He then served as a military policeman at Camp Lejeune, North Carolina, until July 1953, when he was assigned as a mortar unit leader with the 4th Marine Corps Reserve Rifle Company at Rome, Georgia. He was later assigned to the 10th Marines, Camp Lejeune, and retired from the Marine Corps in 1972 as a master sergeant.

Master Sergeant McLaughlin died on January 14, 1977, and was buried in Mount Hebron Cemetery, in Leeds, Alabama.

OLA LEE MIZE

Rank and Organization: Sergeant, 2nd Platoon, Company K, 15th Infantry Regiment, 3rd Infantry Division; **Military Service Branch**: U. S. Army; **Conflict**: Vietnam War; **Medal of Honor Action Date**: June 10, 1953; **Medal of Honor Action Place**: Near Surang-Ni, Korea

Presentation: September 7, 1954, Denver, Colorado, presented by President Dwight Eisenhower

Accredited to: Gadsden, Etowah County, Alabama

Born: August 28, 1931, Albertville, Marshall County, Alabama

Citation

M/Sgt. Mize, a member of Company K, distinguished himself by conspicuous gallantry and outstanding courage above and beyond the call of duty in action against the enemy. Company K was committed to the defense of "Outpost Harry," a strategically valuable position, when the enemy launched a heavy attack. Learning that a comrade on a friendly listening post had been wounded, he moved through the intense barrage, accompanied by a medical aidman, and rescued the wounded soldier. On returning to the main position he established an effective defense system and inflicted heavy casualties against attacks from determined enemy assault forces which had penetrated into trenches within the outpost area. During his fearless actions he was blown down by artillery and grenade blasts three times, but each time he dauntlessly returned to his position, tenaciously fighting and successfully repelling hostile attacks. When enemy onslaughts ceased, he took his few men and moved from bunker to bunker, firing through apertures and throwing grenades at the foe, neutralizing their positions. When an enemy soldier stepped out behind a comrade, prepared to fire, M/Sgt. Mize killed him, saving the life of his fellow soldier. After rejoining the platoon, moving from man to man, distributing ammunition, and shouting words of encouragement he observed a friendly machine-gun position overrun. He immediately fought his way to the position, killing 10 of the enemy and dispersing the remainder. Fighting back to the command post, and finding several friendly wounded there, he took a position to protect them. Later, securing a radio, he directed friendly artillery fire upon the attacking enemy's routes of approach. At dawn he helped regroup for a counterattack which successfully drove the enemy from the outpost. M/Sgt. Mize's valorous conduct and unflinching courage reflect lasting glory upon himself and uphold the noble traditions of the military service.

Epilogue

Mize was born in Albertville, Alabama, the son of a sharecropper. He left school after 9th grade to help support his family. After several years of working for low pay, he attempted to enlist in the Army, but was rejected for being too light, at 120 pounds (54 kg). He tried repeatedly to enlist and was eventually accepted, joining the Army in 1950 from Gadsden, Alabama.

Assigned to the 82nd Airborne Division, Mize planned to finish his term of service and return to school. When the Korean War began, he changed his plans and reenlisted in hopes of seeing combat. He volunteered for a front-line unit and ended up as a sergeant in Company K of the 15th Infantry Regiment, 3rd Infantry Division. On June 10, 1953, his unit was manning Outpost Harry near Surang-ni, Korea, when the post came under heavy enemy attack. Mize organized defensive positions, rescued wounded soldiers, and engaged the enemy until reinforcements arrived about noon the next day. He was subsequently promoted to master sergeant and, on September 24, 1954, awarded the Medal of Honor for his actions at Outpost Harry.

Mize later joined the Special Forces where he gained a commission and served three tours of duty in the Vietnam War. In 1965, he was assigned to the Special Forces Training Group, where he was the Advanced Training Committee chief for SCUBA, HALO, and the SKY HOOK schools. Colonel Mize is credited for being the officer responsible for starting the present day Combat Divers Qualification Course in Key West, FL. From 1966 to 1967, he was again assigned to Vietnam with the 5th Special Forces Group where he was an Operational A, B, and C Detachment Commander and once more in 1969, where he was assigned to the 5th Special Forces Group as the Commander of the 3rd Mobile Strike Force Command (Cambodian Troops). During his tour while commander of B-36 3rd Mobile Strike Force he was awarded the Silver Star for Valor in 1969. Going on to meet presidents Eisenhower and Kennedy. He retired in 1981 as a colonel.

Mize died of cancer on March 12, 2014 at Gadsden, Alabama, aged 82.

Vietnam War Recipients of the Medal of Honor

ROBERT L. HOWARD

Rank and Organization: First Lieutenant (Rank at the time of action: Sergeant First Class), 5th Special Forces Group (Airborne), 1st Special Forces; **Military Service Branch**: U. S. Army; **Conflict**: Vietnam War; **Medal of Honor Action Date**: December 30, 1968; **Medal of Honor Action Place**: Republic of Vietnam

Presentation: March 2, 1971 at The White House, presented by President Richard Nixon

Accredited to: Montgomery, Montgomery County, Alabama

Born: July 11, 1939, Opelika, Lee County, Alabama

Died: December 23, 2009, Waco, Texas

Buried: Arlington National Cemetery, Arlington, Virginia

Citation

For conspicuous gallantry and intrepidity in action at the risk of his life above and beyond the call of duty. 1st Lt. Howard (then Sfc.) distinguished himself while serving as platoon sergeant of an American-Vietnamese platoon which was on a mission to rescue a missing American soldier in enemy-controlled territory in the Republic of Vietnam. The platoon had left its helicopter landing zone and was moving out on its mission when it was attacked by an estimated two-company force. During the initial engagement, 1st Lt. Howard was wounded and his weapon destroyed by a grenade explosion. 1st Lt. Howard saw his platoon leader had been wounded seriously and was exposed to fire. Although unable to walk, and weaponless, 1st Lt. Howard unhesitatingly crawled through a hail of fire to retrieve his wounded leader. As 1st Lt. Howard was administering first aid and removing the officer's equipment, an enemy bullet struck one of the ammunition pouches on the lieutenant's belt, detonating several magazines of ammunition. 1st Lt. Howard momentarily sought cover and then realizing that he must rejoin the platoon, which had been disorganized by the enemy attack, he again began dragging the seriously wounded officer toward the platoon area. Through his outstanding example of indomitable courage and bravery, 1st Lt. Howard was able to rally the platoon into an organized defense force. With complete disregard for his safely, 1st Lt. Howard crawled from position to position, administering first aid to the wounded, giving encouragement to the defenders and directing their fire on the encircling enemy. For 3 and one half hours 1st Lt. Howard's small force and supporting aircraft successfully repulsed enemy attacks and finally were in sufficient control to permit the landing of rescue helicopters. 1st Lt. Howard personally supervised the loading of his men and did not leave the bullet-swept landing zone until all were aboard safely. 1st Lt. Howard's gallantry in action, his complete devotion to the welfare of his men at the risk of his life were in keeping with the highest traditions of the military service and reflect great credit on himself, his unit, and the U.S. Army.

Epilogue

Howard enlisted in the Army in 1956 at Montgomery, Alabama and retired as colonel, Army Special Forces, in 1992.

Howard's service in Vietnam included assignments with 1/327th Airborne Infantry, 1st Brigade, 101st Airborne Division, 5th Special Forces Group and MACV-SOG

As a staff sergeant of the highly classified Military Assistance Command, Vietnam – Studies and Observations Group (MACV-SOG), Howard was recommended for the Medal of Honor on three occasions for three individual actions during thirteen months spanning 1967–1968. The first two nominations were downgraded to a Silver Star and the Distinguished Service Cross due to the covert and top secret nature of the operations in which Howard participated.

As a sergeant first class of the SOG, he risked his life during a rescue mission in Cambodia on December 30, 1968, while second in command of a platoon-sized Hatchet Force that was searching for missing American soldier Robert Scherdin for which he was awarded the Medal of Honor. He learned of the award over a two-way radio while under enemy fire, immediately after being wounded, resulting in one of his eight Purple Hearts.

Howard was wounded 14 times during a 54-month period in the Vietnam War. For his distinguished service, Howard received a direct appointment from Master Sergeant to First Lieutenant in December 1969.

Howard graduated from Ranger School class 7-73 in May 1973 and served with the 2nd Ranger Battalion at Fort Lewis, Washington as company commander. From 1977 to 1978 he served as Mountain Ranger Training instructor.

Howard later served as officer-in-charge of Special Forces training at Camp Mackall, near Ft. Bragg, N.C., and later, commanding the Mountain Ranger Training Camp at Dahlonega, Georgia

Howard graduated from the National War College, Class 1987–1988.

He received two master's degrees during his Army career which spanned 1956 to 1992. Howard retired as a colonel in 1992.

He was one of the most decorated soldiers in the Vietnam War. *NBC News* said that Howard may have been the most highly decorated American soldier of the modern era, while KWTX-TV states that he was "said to be the most decorated service member in the history of the United States". John Plaster in his 1998 book *SOG: The Secret Wars of America's Commandos in Vietnam* states that Howard "remains to this day the most highly decorated American soldier."

His residence was in Texas and he spent much of his free time working with veterans until the time of his death. He also took periodic trips to Iraq to visit active duty troops.

Howard died of pancreatic cancer at a hospital in Waco, Texas, on December 23, 2009. He was survived by four children and five grandchildren.

His funeral was in Arlington National Cemetery on February 22, 2010.

RODNEY JOHN EVANS

POSTHUMOUS AWARD

Rank and Organization: Sergeant, Company D, 1st Battalion, 12th cavalry, 1st cavalry Division; **Military Service Branch**: U. S. Army; **Conflict**: Vietnam War; **Medal of Honor Action Date**: July 18, 1969; **Medal of Honor Action Place**: Tay Ninh Province, Republic of Vietnam

Presentation: April 22, 1971 at The White House, presented by President Richard Nixon

Accredited to: Montgomery, Montgomery County, Alabama

Born: July 17, 1948, Chelsea, Suffolk County, Massachusetts

Died: July 18, 1969, Republic of Vietnam

Buried: Liberty Cemetery (MH), Florala, Alabama

Citation

For conspicuous gallantry and intrepidity in action at the risk of his life above and beyond the call of duty. Sgt. Evans distinguished himself by extraordinary heroism while serving as a squad leader in a reconnaissance sweep through heavy vegetation to reconnoiter a strong enemy position. As the force approached a well-defined trail, the platoon scout warned that the trail was booby-trapped. Sgt. Evans led his squad on a route parallel to the trail. The force had started to move forward when a nearby squad was hit by the blast of a concealed mine. Looking to his right Sgt. Evans saw a second enemy device. With complete disregard for his safety he shouted a warning to his men, dived to the ground, and crawled toward the mine. Just as he reached it an enemy soldier detonated the explosive and Sgt. Evans absorbed the full impact with his body. His gallant and selfless action saved his comrades from probable death or injury and served as an inspiration to his entire unit. Sgt. Evans' gallantry in action at the cost of his life was in keeping with the highest traditions of the military service and reflect great credit upon himself, his unit, and the U.S. Army.

Epilogue

Evans was the adopted son of the Evans family of Florala, Alabama. His parents owned a dry cleaning business. After joining the US Army, he married his high school sweetheart, Barbara Geohagan. Evans was then sent to Vietnam and successfully completed a full tour of duty.

Upon his return to the United States, he was stationed at Fort Rucker, Alabama, not far from where he grew up. Due to financial constraints his wife lived with her parents and would drive up to Fort Rucker on weekends to pick him up and take him home.

On Thursday, September 5, 1968, Barbara was killed in a car accident after going to Ft Rucker to take care of some business. She lost control of her car. Evans became very depressed, re-enlisted and volunteered for another tour in Vietnam. It took over a year for his papers to be processed and for Evans to receive orders for his return to Vietnam. By then, with the help of close friends he had come to grips with his loss. Shortly before his return to Vietnam, Rodney told friends and his twin brother, Wyman, he no longer wanted to die and regretted re-upping for Vietnam. After he had left for Vietnam, the Dunn family (with whom he and Wyman lived after Army enlistment and again after Barbara's death) petitioned legislators and the Army to cut-short Evans' tour of duty due to the reasons he re-upped—that he was under mental duress. The Dunn's received a call from Evans shortly before his death, while he was on a short "rest and relaxation" break. He was so grateful for the efforts. The word came from one legislator that Evans' paperwork was being processed and, indeed, he would be returned. It would take a few weeks, but he would be released from the commitment. Less than a week later, Evans was killed in action during Operation Toan Thang III.

Rodney Evans died one day after his twenty-first birthday and is buried in Liberty Hill Cemetery, Florala, Alabama. His name can be found on Panel 20W - Row 014 of the Vietnam Memorial Wall in Washington, D.C.

MATTHEW LEONARD

<u>POSTHUMOUS AWARD</u>

Rank and Organization: Sergeant First Class, Company B, 1st Battalion, 16th Infantry, 1st Infantry Division; **Military Service Branch**: U. S. Army; **Conflict**: Vietnam War; **Medal of Honor Action Date**: February 28, 1967; **Medal of Honor Action Place**: Republic of Vietnam

Presentation: December 19, 1968, The Pentagon, presented to his family by Army Secretary Stanley R. Resor

Accredited to: Birmingham, Jefferson County, Alabama

Born: November 26, 1929, Eutaw, Greene County, Alabama

Died: February 28, 1967, Republic of Vietnam

Buried: Fort Mitchell National Cemetery (MH) (14-27), remains removed from Shadow Lawn Cemetery (MH) (4-45), Fort Mitchell, Alabama

Citation

Citation: For conspicuous gallantry and intrepidity in action at the risk of his life above and beyond the call of duty. His platoon was suddenly attacked by a large enemy force employing small arms, automatic weapons, and hand grenades. Although the platoon leader and several other key leaders were among the first wounded, P/Sgt. Leonard quickly rallied his men to throw back the initial enemy assaults. During the short pause that followed, he organized a defensive perimeter, redistributed ammunition, and inspired his comrades through his forceful leadership and words of encouragement. Noticing a wounded companion outside the perimeter, he dragged the man to safety but was struck by a sniper's bullet which shattered his left hand. Refusing medical attention and continuously exposing himself to the increasing fire as the enemy again assaulted the perimeter, P/Sgt. Leonard moved from position to position to direct the fire of his men against the well camouflaged foe. Under the cover of the main attack, the enemy moved a machine gun into a location where it could sweep the entire perimeter. This threat was magnified when the platoon machine gun in this area malfunctioned. P/Sgt. Leonard quickly crawled to the gun position and was helping to clear the malfunction when the gunner and other men in the vicinity were wounded by fire from the enemy machine gun. P/Sgt. Leonard rose to his feet, charged the enemy gun and destroyed the hostile crew despite being hit several times by enemy fire. He moved to a tree, propped himself against it, and continued to engage the enemy until he succumbed to his many wounds. His fighting spirit, heroic leadership, and valiant acts inspired the remaining members of his platoon to hold back the enemy until assistance arrived. P/Sgt. Leonard's profound courage and devotion to his men are in keeping with the highest traditions of the military

service, and his gallant actions reflect great credit upon himself and the U.S. Army.

Epilogue

Leonard was born and raised in Alabama and graduated from A P. Parker High School before entering the Army in 1947. Before Vietnam, he served in combat during the Korean War from 1951-1953. At the time of his death, he was survived by his wife, Lois, and five children: Carl (21), Lavon (17), Brenda (16), Wanda (13), and Paula (8). The Medal was presented to his wife by Army Secretary Resor at a Pentagon ceremony. Leonard is buried at Fort Mitchell National Cemetery, Fort Mitchell, Alabama.

DON L. MICHAELS

POSTHUMOUS AWARD

Rank and Organization: Specialist Fourth Class, Company C, 4th Battalion, 503rd Infantry, 173rd Airborne Brigade; **Military Service**

Branch: U. S. Army; **Conflict**: Vietnam War; **Medal of Honor Action Date**: April 6, 1967; **Medal of Honor Action Place**: Republic of Vietnam

Presentation: May 23, 1969 at The White House, presented by President Richard Nixon

Accredited to: Montgomery, Montgomery County, Alabama

Born: July 31, 1947, Florence, Lauderdale County, Mississippi

Died: April 8, 1967, Republic of Vietnam

Buried: Mount Pleasant Baptist Church (MH), Lexington, Alabama

Citation

For conspicuous gallantry and intrepidity at the risk of his life above and beyond the call of duty. Sp4c. Michael, U.S. Army, distinguished himself while serving with Company C. Sp4c. Michael was part of a platoon which was moving through an area of suspected enemy activity. While the rest of the platoon stopped to provide security, the squad to which Sp4c. Michael was assigned moved forward to investigate signs of recent enemy activity. After moving approximately 125 meters, the squad encountered a single Viet Cong soldier. When he was fired upon by the squad's machine gunner, other Viet Cong opened fire with automatic weapons from a well-concealed bunker to the squad's right front. The volume of enemy fire was so withering as to pin down the entire squad and halt all forward movement. Realizing the gravity of the situation, Sp4c. Michael exposed himself to throw two grenades, but failed to eliminate the enemy position. From his position on the left flank, Sp4c. Michael maneuvered forward with two more grenades until he was within 20 meters of the enemy bunkers, when he again exposed himself to throw two grenades, which failed to detonate. Undaunted, Sp4c. Michael made his way back to the friendly positions to obtain more grenades. With two grenades in hand, he again started his perilous move towards the enemy bunker, which by this time was under intense artillery fire from friendly positions. As he neared the bunker, an enemy soldier attacked him from a concealed position. Sp4c. Michael killed him with his rifle and, in spite of the enemy fire and the exploding artillery rounds, was successful in destroying the enemy positions. Sp4c. Michael took up pursuit of the remnants of the retreating enemy. When his comrades reached Sp4c. Michael, he had been mortally wounded. His inspiring display of determination and

courage saved the lives of many of his comrades and successfully eliminated a destructive enemy force. Sp4c. Michael's actions were in keeping with the highest traditions of the military service and reflect the utmost credit upon himself and the U.S. Army.

Epilogue

Michael joined the Army from Montgomery, Alabama in 1966, and by April 8, 1967, was serving as a Specialist Four in Company C, 4th Battalion, 503d Infantry, 173d Airborne Brigade. On that day, in the Republic of Vietnam, he single-handedly destroyed a Viet Cong bunker and was then mortally wounded while chasing the retreating enemy soldiers. For his actions, he was posthumously awarded the Medal of Honor.

Michael, aged 19 at his death, was buried at Mount Pleasant Baptist Church, Lexington, Alabama.

GARY "MIKE" ROSE

Rank and Organization: Sergeant, Command and Control Center, Special Operations Augmentation, Command and Control Center, 5th Special Group, 1st Special Forces; **Military Service Branch**: U. S. Army; **Conflict**: Vietnam War; **Medal of Honor Action Date**: September 11-14, 1970; **Medal of Honor Action Place**: Laos

Presentation: October 23, 2017 at The White House, presented by President Donald trump

Accredited to: Panarama City, Los Angeles County, California

Born: October 17, 1947, Watertown, Jefferson County, New York

Citation:

Sergeant Gary Michael Rose distinguished himself by acts of gallantry and intrepidity at the risk of his life above and beyond the call of duty while serving as a Special Forces Medic with a company sized exploitation force, 5th Special Forces Group, 1st Special Forces. Between 11 and 14 September 1970, Sergeant Rose's company was continuously engaged by a well-armed and numerically superior hostile force deep in enemy-controlled territory. Enemy B-40 rockets and mortar rounds rained down while the adversary sprayed the area with small arms and machine gun fire, wounding many and forcing everyone to seek cover. Sergeant Rose, braving the hail of bullets, sprinted fifty meters to a wounded soldier's side. He then used his own body to protect the casualty from further injury while treating his wounds. After stabilizing the casualty, Sergeant Rose carried him through the bullet-ridden combat zone to protective cover. As the enemy accelerated the attack, Sergeant Rose continuously exposed himself to intense fire as he fearlessly moved from casualty to casualty, administering life-saving aid. A B-40 rocket impacted just meters from Sergeant Rose, knocking him from his feet and injuring his head, hand, and foot. Ignoring his wounds, Sergeant Rose struggled to his feet and continued to render aid to the other injured soldiers. During an attempted medevac, Sergeant Rose again exposed himself to enemy fire as he attempted to hoist wounded personnel up to the hovering helicopter, which was unable to land due to unsuitable terrain. The medevac mission was aborted due to intense enemy fire and the helicopter crashed a few miles away due to the enemy fire sustained during the attempted extraction. Over the next two days, Sergeant Rose continued to expose himself to enemy fire in order to treat the wounded, estimated to be half of the company's personnel. On September 14, during the company's eventual helicopter extraction, the enemy launched a full-scale offensive. Sergeant Rose, after loading wounded personnel on the first set of extraction helicopters, returned to the outer perimeter under enemy fire, carrying friendly casualties and moving wounded personnel to more secure positions until they could be evacuated. He then returned to the perimeter to help repel the enemy until the final extraction helicopter arrived. As the final helicopter was loaded, the enemy began to overrun the company's position, and the helicopter's Marine door gunner was shot in the neck. Sergeant Rose instantly administered critical medical treatment onboard the helicopter, saving the Marine's life. The helicopter carrying Sergeant Rose crashed several hundred meters from the evacuation point, further injuring Sergeant Rose and the personnel on board. Despite his numerous wounds from the past three days, Sergeant Rose continued to pull and carry unconscious and wounded personnel out of the burning wreckage and continued to administer aid to the wounded until another extraction helicopter arrived. Sergeant Rose's extraordinary heroism and selflessness above and beyond the call of duty were critical to saving numerous lives over that four day time period. His actions are in keeping with the highest traditions of military service and reflect great credit upon himself, the 1st Special Forces, and the United States Army.

Epilogue

Rose was born on 17 October 1947 in Watertown, New York, and later moved to the Los Angeles area. In 1965, he graduated from the James Monroe High School in Northridge. On 4 April 1967, he volunteered for the United States Army to avoid being drafted into the Marine Corps after his father, who had been in the Marine Corps during World War II, suggested

Rose attended basic training at Fort Ord and Infantry Advanced Individual Training at Fort Gordon. He was promoted to private first class after graduating from the latter and was sent to the United States Army Airborne School due to his high aptitude test scores. Rose entered Special Forces training at Fort Bragg in October, graduating a year later as a Special Forces medic. Rose was first assigned to the 7th Special Forces Group (Airborne). He reenlisted to be able to choose where he wanted to serve and in April 1969 was assigned to the 46th Special Forces Company in Lopburi, Thailand, where he trained Thai soldiers and border police medics.

In April 1970, Rose requested transfer to South Vietnam and was assigned to Military Assistance Command, Vietnam – Studies and Observations Group (SOG), for which the 5th Special Forces Group (Airborne) provided administrative support. He was based at Forward Operating Base II at Kontum, where he treated the wounded and local civilians. Rose was wounded on his first mission in June 1970, receiving his first Purple Heart and Bronze Star Medal. On 11 September, he and a company-size exploitation force of Americans,

Vietnamese, and Montagnards were inserted by CH-53 Sea Stallion helicopters 70 kilometers inside Laos near Chavane in Operation Tailwind, a diversionary operation. Then-Sergeant Rose was responsible for medical care for fifteen other

After being inserted, the force advanced deeper into enemy territory, soon making contact with a PAVN squad, wounding two Americans and two Montagnards, one of whom was trapped outside the unit's defensive perimeter. Rose engaged the enemy while treating and stabilizing the soldier, carrying him through heavy fire back to the defensive positions. The company continued moving deeper into Laos following a PAVN withdrawal, engaging more PAVN forces and suffering more casualties. Disregarding his personal safety, Rose treated wounded under heavy fire, engaging the PAVN to reach the wounded.

The fire became so intense Rose was forced to crawl from position to position in order to treat the wounded, giving words of encouragement and directing the fire of the inexperienced and terrified South Vietnamese and Montagnard troops. Over the next few days, the company marched west and deeper into Laos, fending off incessant attacks from elements ranging from squad-sized to company-sized, covered by Air Force gunships. On 12 September, during an assault by a PAVN company-sized element, one of the Montagnards was wounded 40 to 50 meters outside the perimeter. Dodging enemy fire, Rose ran and crawled his way to the wounded man, shielding him with his own body while he treated him. He dragged the Montagnard back to the company with one hand while firing at the PAVN with his gun in his other hand.

When Rose returned to the company perimeter with the wounded soldier, he was sprayed with shrapnel from a B-40 rocket propelled grenade (RPG) in his back and leg, severely crippling his foot. For the rest of the mission, he used a stick as a crutch, continuing to treat the wounded while ignoring his own wounds. A MEDEVAC helicopter was later called in to evacuate the company's many wounded, but was unable to land in the small opening where the company was located. Rose stood up, fully exposed to the heavy enemy fire, and attempted to pass the wounded up to the hovering helicopter's crew. However, the intensity of the fire forced the pilot to abort the mission, and the helicopter, severely damaged, crashed a few kilometers away.

With the aid of close air support, the company was able to break out of their defensive position and Rose improvised litters for the wounded, who now were over half the company. Despite his own wounds, he never took time to eat, rest or care for his own wounds while treating the other soldiers. On the night of 13–14 September, the company was surrounded by the PAVN in their position. Rose dug trenches for the wounded and treated their injuries, as the enemy spent the night bombarding the company with rockets, grenades, and mortars. Moving from position to position, he exposed himself to PAVN fire, encouraging the soldiers and treating the wounded.

On 14 September, the last day of the operation, after destroying a PAVN base camp, the company was informed by a Forward Air Controller (FAC) that over 500 PAVN were advancing on their position. Supported by gunships, the company moved to a landing zone, setting up a perimeter while each platoon boarded the helicopters as the PAVN attacked from all sides, inflicting even more casualties. Rose exposed himself again under heavy fire, retrieving the dead and wounded and bringing them back to the perimeter. In great pain, he continued to treat the wounded, disregarding his own safety. When the extraction helicopters arrived, he returned to the outer perimeter to help engage the PAVN and repel the assault. Rose boarded the last extraction helicopter, delivering aimed accurate fire at the PAVN only an estimated 50 meters away as he hobbled up the loading ramp.

Soon after the helicopter lifted off, it was hit by anti-aircraft rounds, and its engine stopped at an altitude of 4,500 feet (1,400 m). Rose was notified that a Marine door gunner had been shot through the neck and he rendered lifesaving medical treatment to the man before the helicopter crashed, several kilometers away from the original extraction point. Rose was thrown from the helicopter before it crashed, and while still dazed and wounded crawled back into the wreckage to pull out wounded and unconscious soldiers, knowing that it could explode at any moment. He continued to treat the wounded until another helicopter arrived to extract them. Upon returning to base, Rose, covered in blood and wounds, refused all treatment until the other wounded were treated first. He was credited with treating between 60 and 70 wounded and saving many lives; only three Montagnards died in Operation Tailwind.

Rose was nominated for the Medal of Honor, but the award was downgraded to the Distinguished Service Cross due to the classified nature of the mission in Laos. MACV commander General Creighton Abrams formally presented Rose the Distinguished Service Cross on 16 January 1971.

In April, when his tour in Vietnam ended, Rose was sent to the Spanish Language School in Anacostia, Washington, D.C. At the school Rose made the decision to attend Officer Candidate School because extending his contract with the Army would enable him to bring his new wife, Margaret, with him to Panama. After completing the school, he was assigned to the 8th Special Forces Group in Panama until August 1973, when he was selected to attend Officer Candidate School at Fort Benning. Rose received his commission as a second lieutenant in the Field Artillery in December, and attended the Field Artillery Officer Basic Course at Fort Sill. He graduated from Cameron University in Lawton with a Bachelor of Arts in Education and Military Science in December 1977. In 1978, Rose attended the Field Artillery Officer Advanced Course, and then was posted to various successive field assignments in Germany, New Mexico, South Korea, and Fort Sill over the next years. He retired from the Army with the rank of captain in May 1987.

Rose attended the University of Oklahoma, graduating in December 1989 with a Master of Arts in Communication. He worked as a writer of operator, user, and maintenance manuals and training designer for the manufacturing industry before finally retiring in 2010. Rose worked at Raytheon, which he left in 2003. He moved to Huntsville in 2005 to live near a friend who was a fellow Vietnam War veteran. In retirement, Rose remains involved in charity work, mainly through the Knights of Columbus.

In 1998, a joint CNN and *Time* magazine report incorrectly described Operation Tailwind as a mission to kill American defectors and claimed that American troops used sarin gas on civilians. Rose was among the Tailwind veterans called to the Pentagon in late June for interviews on the operation, and they refuted the claims. The story was retracted after a Department of Defense report concluded that the story was incorrect. After the controversy, an initiative began by SOG veterans to get their comrades' heroism in Operation Tailwind recognized. In 2013, Rose received a call from Eugene McCarley, who commanded the company in Operation Tailwind, who told him that Army veteran and SOG researcher Neil Thorne wanted to request an upgrade to the Medal of Honor for Rose's DSC. Thorne requested information from Rose, and then-Secretary of Defense Ash Carter approved the award upgrade in 2016, and Alabama Representative Mo Brooks and Senator Jeff Sessions wrote Rose's name into the National Defense Authorization Act for Fiscal Year 2017, waiving the requirement the medal must be awarded within five years of the action. On 20 September 2017, the White House announced President Donald Trump would present Rose the Medal of Honor on 23 October. On 23 October, President Donald J. Trump presented Rose the Medal of Honor in a White House ceremony.

Rose married Margaret in 1971 and has two daughters: Claire Rose and Sarah Bowen, and a son, Michael.

JAMES M. SPRAYBERRY

Rank and Organization: First Lieutenant. Company D, 5th Battalion, 7th Cavalry, 1st Cavalry Division; **Military Service Branch**: U. S. Army; **Conflict**: Vietnam War; **Medal of Honor Action Date**: September April 25, 1968; **Medal of Honor Action Place**: Republic of Vietnam

Presentation: October 9, 1969 at The White House, presented by President Richard Nixon

Accredited to: Montgomery, Montgomery County, Alabama

Born: April 24, 1947, Troup County, Georgia

Citation:

For conspicuous gallantry and intrepidity in action at the risk of his life above and beyond the call of duty. Capt. Sprayberry, Armor, U.S. Army, distinguished himself by exceptional bravery while serving as executive officer of Company D. His company commander and a great number of the men were wounded and separated from the main body of the company. A daylight attempt to rescue them was driven back by the well-entrenched enemy's heavy fire. Capt. Sprayberry then organized and led a volunteer night patrol to eliminate the intervening enemy bunkers and to relieve the surrounded element. The patrol soon began receiving enemy machine-gun fire. Capt. Sprayberry quickly moved the men to protective cover and without regard for his own safety, crawled within close range of the bunker from which the fire was coming. He silenced the machine gun with a hand grenade. Identifying several one-man enemy positions nearby, Capt. Sprayberry immediately attacked them with the rest of his grenades. He crawled back for more grenades and when two grenades were thrown at his men from a position to the front, Capt. Sprayberry, without hesitation, again exposed himself and charged the enemy-held bunker, killing its occupants with a grenade. Placing two men to cover his advance, he crawled forward and neutralized three more bunkers with grenades. Immediately thereafter, Capt. Sprayberry was surprised by an enemy soldier who charged from a concealed position. He killed the soldier with his pistol and with continuing disregard for the danger neutralized another enemy emplacement. Capt. Sprayberry then established radio contact with the isolated men, directing them toward his position. When the two elements made contact he organized his men into litter parties to evacuate the wounded. As the evacuation was nearing completion, he observed an enemy machine-gun position which he silenced with a grenade. Capt. Sprayberry returned to the rescue party, established security, and moved to friendly lines with the wounded. This rescue operation, which lasted approximately 7 1/2 hours, saved the lives of many of his fellow soldiers. Capt. Sprayberry personally killed 12 enemy soldiers, eliminated two machine guns, and destroyed numerous enemy bunkers. Capt. Sprayberry's indomitable spirit and gallant action at great personal risk to his life are in keeping with the highest traditions of the military service and reflect great credit upon himself, his unit, and the U.S. Army.

Epilogue

Born in LaGrange, Georgia, on April 24, 1947, Sprayberry grew up in Sylacauga, Alabama. He joined the Army from Montgomery in 1967, and by April 25, 1968 was serving as a first lieutenant in Company D, 5th Battalion, 7th Cavalry Regiment, 1st Cavalry Division (Airmobile). On that day, in the Republic of Vietnam, he led a patrol which rescued men who had been wounded and cut off from the rest of the company. In the course of the rescue mission, Sprayberry personally destroyed several enemy bunkers and machine gun emplacements. He was subsequently promoted to captain and awarded the Medal of Honor for his actions.

Sprayberry reached the rank of lieutenant colonel before retiring from the army in 1988.

LEO KEITH THORSNESS

Rank and Organization: Major, 357th Tactical Fighter Squadron, 355th Tactical Fighter Wing; **Military Service Branch**: U. S. Air Force; **Conflict**: Vietnam War; **Medal of Honor Action Date**: April 19, 1967; **Medal of Honor Action Place**: Over North Vietnam

Presentation: October 15, 1973 at The White House, presented by President Richard Nixon

Accredited to: Walnut Grove, Redwood County, Minnesota

Born: February 14, 1932, Walnut Grove, Redwood County, Minnesota

Died: May 2, 2017, St. Augustine, Florida

Buried: Arlington National Cemetery, Arlington, Virginia

Citation:

For conspicuous gallantry and intrepidity in action at the risk of his life above and beyond the call of duty. As pilot of an F-105 aircraft, Lt. Col. Thorsness was on a surface-to-air missile suppression mission over North Vietnam. Lt. Col. Thorsness and his wingman attacked and silenced a surface-to-air missile site with air-to-ground missiles, and then destroyed a second surface-to-air missile site with bombs. In the attack on the second missile site, Lt. Col. Thorsness' wingman was shot down by intensive antiaircraft fire, and the two crewmembers abandoned their aircraft. Lt. Col. Thorsness circled the descending parachutes to keep the crewmembers in sight and relay their position to the Search and Rescue Center. During this maneuver, a MIG-17 was sighted in the area. Lt. Col. Thorsness immediately initiated an attack and destroyed the MIG. Because his aircraft was low on fuel, he was forced to depart the area in search of a tanker. Upon being advised that two helicopters were orbiting over the downed crew's position and that there were hostile MIGs in the area posing a serious threat to the helicopters, Lt. Col. Thorsness, despite his low fuel condition, decided to return alone through a hostile environment of surface-to-air missile and antiaircraft defenses to the downed crew's position. As he approached the area, he spotted four MIG-17 aircraft and immediately initiated an attack on the MIGs, damaging one and driving the others away from the rescue scene. When it became apparent that an aircraft in the area was critically low on fuel and the crew would have to abandon the aircraft unless they could reach a tanker, Lt. Col. Thorsness, although critically short on fuel himself, helped to avert further possible loss of life and a friendly aircraft by recovering at a forward operating base, thus allowing the aircraft in emergency fuel condition to refuel safely. Lt. Col. Thorsness' extraordinary heroism, self-sacrifice, and personal bravery involving conspicuous risk of life were in the highest traditions of the military service, and have reflected great credit upon himself and the U.S. Air Force.

Epilogue

Thorsness was born February 14, 1932, in Walnut Grove, Minnesota, where his family had a farm. There, he earned the Eagle Scout award from the Boy Scouts of America. He is one of only nine known Eagle Scouts who also received the Medal of Honor. The others are Aquilla J. Dyess and Mitchell Paige of the U.S. Marine Corps, Robert Edward Femoyer and Jay Zeamer Jr. of the U.S. Army Air Forces, Arlo L. Olson, Benjamin L. Salomon of the United States Army, and Eugene B. Fluckey and Thomas R. Norris of the United States Navy. In 2010, Thorsness received the Distinguished Eagle Scout Award.

He attended South Dakota State College in Brookings, South Dakota, where he met his future wife, Gaylee Anderson, also a freshman. They married in 1953 and had a daughter, Dawn.

Thorsness enlisted in the Air Force in 1951 at the age of 19 because his brother was then serving in the Korean War. In 1954, he received his commission as an officer and his wings with a rating of pilot through the USAF Aviation Cadet program in Class 54-G. He later earned a bachelor's degree from the University of Omaha in 1964, and a master's degree in Defense Systems Management from the University of Southern California. His initial assignment was as a pilot in the Strategic Air Command and Tactical Air Command, but he completed training as a fighter pilot and flew both F-84 and F-100 jets before transitioning to the F-105 Thunderchief.

In the autumn of 1966, after completing F-105 "Wild Weasel" training at George AFB, California, he was assigned to the 355th Tactical Fighter Wing based at Takhli Royal Thai Air Force Base in Thailand, flying as aircraft commander in F-105F's, tasked with locating and destroying North Vietnamese surface-to-air missile (SAM) sites.

On April 19, 1967, Major Thorsness and his Electronic Warfare Officer, Captain Harold E. Johnson, flying F-105F AF Ser. No. 63-8301, led Kingfish flight (three F-105F Weasel aircraft and an F-105D single-seater) on a Wild Weasel SAM suppression mission. The strike force target was JCS target 22.00, the Xuan Mai army training compound, near heavily defended Hanoi. Thorsness directed Kingfish 03 and 04, the second element of F-105s, to troll north while he and his wingman maneuvered south, forcing defending gunners to divide their attention. Thorsness located two SAM sites and fired a Shrike missile to attack one, whose radar went off the air. He destroyed the second with cluster bombs, scoring a direct hit.

After this initial success, matters turned for the worse. Kingfish 02, crewed by Majors Thomas M. Madison and Thomas J. Sterling, flying aircraft F-105F AF Ser. No. 63-8341, was hit by anti-aircraft fire and both crewmen had to eject. Unknown to Thorsness, Kingfish 03 and 04 had been attacked by MiG-17s flying a low-altitude wagon wheel defensive formation. The afterburner of one of the F-105s wouldn't light and the element had disengaged and returned to base, leaving Kingfish 01 to fight solo.

As their F-105 circled the parachutes of Kingfish 02-Alpha and 02-Bravo, relaying the position to Crown, the airborne search and rescue HC-130 command aircraft, Johnson spotted a MiG-17 off their right wing. 8301, though not designed for air-to-air combat, responded well as Thorsness attacked the MiG and destroyed it with 20-mm cannon fire, just as a second MiG closed on his tail. Low on fuel, Thorsness outran his pursuers and left the battle area to rendezvous with a KC-135 tanker over Laos.

Thorsness described the incident:

It appeared the MiG was going after the chutes so I took off after him. I was a little high, dropped down to about 1000 feet, and headed north after him. We were doing about 550 knots and really catching up fast. At about 3000 feet (range) I fired a burst but missed. I lined him up again and was closing very fast. I was a bit below him now. At 700 feet or so I pull my trigger and pulled the pipper through him. Parts of his left started coming off. Suddenly I realized that Harry Johnson was frantically trying to get my attention. There were a couple of MiGs on our tail! If I had hit that MiG dead on, we probably would have swallowed some of his debris. But we got him! I lit the burner, dropped down as low as possible, and ducked into the hills west of Hanoi. The MiGs could not keep up with us.

As this occurred, the initial element of the rescue force—a pair of A-1E "Sandies"—arrived to locate the position of the downed crewmen before calling in the waiting HH-53 Jolly Green helicopters orbiting at a holding point over Laos. Thorsness, with only 500 rounds of ammunition left, turned back from the tanker to fly RESCAP (rescue combat air patrol) for the Sandies and update them on the situation and terrain. As Thorsness approached the area, briefing the Sandies, he spotted MiG-17s in a wagon wheel orbit around him and attacked, probably destroying another that flew across his path.

He commented:

One of the MiGs flew right into my gunsight at about 2000 feet (range). I pulled the trigger and saw pieces start falling off the aircraft. They hadn't seen us, but they did now! Johnson shouted at me that we had four more MiGs on our tail and they were closing fast. I dropped down on the deck, sometimes as low as fifty feet, hit the burner, and twisted through the hills and valleys trying to lose them.

Pairs of MiGs attacked each propeller-driven Sandy as it came out of its turn in search orbit, shooting down the leader (Maj. John S. Hamilton in A-1E 52-133905) with cannon fire when he failed to heed warnings from Sandy 02 to break into the attack, and forced the wingman into a series of repeated evasive turns. Sandy 02 reported the situation and Thorsness advised him to keep turning and announced his return.

Although all of his ammunition had been depleted, Thorsness reversed and flew back to the scene, hoping in some way to draw the MiGs away from the surviving A-1. However, as he re-engaged, Panda flight from the 355th TFW strike force arrived back in the area. It had dropped its ordnance on the target and was en route to its post-strike aerial refueling when Kingfish 02 went down. Panda had jettisoned its wing tanks, making the rescue radar controller reluctant to use it to CAP the rescue effort, but it filled its internal tanks and returned to North Vietnam at high altitude to conserve fuel.

Panda's four F-105s burst through the defensive circle at high speed, then engaged the MiGs in a turning dogfight, permitting Kingfish 01 to depart the area after a 50-minute engagement against SAMs, antiaircraft guns, and MiGs. Panda 01 (Capt William E. Eskew) shot down a MiG, during which the surviving Sandy escaped, and he and his wingman Panda 02 (Capt Paul A. Seymour) each damaged one of the others. Two other MiGs were shot down by members of a third F-105 strike flight, Nitro 01 (Major Jack W. Hunt) and Nitro 03 (Maj. Theodore G. "Ted" Tolman), in another of the 17 MiG engagements on this mission.

Again low on fuel and facing nightfall, Thorsness was headed towards a tanker when Panda 03 (Capt Howard L. Bodenhamer), an F-105 of the flight that had rescued Sandy 02, transmitted by radio that he was critically low on fuel. Thorsness quickly calculated that Kingfish 01 had sufficient fuel to fly to Udorn, near the Mekong River and 200 miles closer, so he vectored the tanker toward Panda 03. When within 60 miles of Udorn, he throttled back to idle and "glided" toward the base, touching down "long" (mid-runway) as his fuel totalizer indicated empty tanks.

The mission was recreated by The History Channel as part of Episode 12 ("Long Odds") of its series *Dogfights*, and first telecast on January 19, 2007. However, Kingfish Flight was incorrectly referred to as "Cadillac" flight.

On April 30, 1967, on their 93rd mission (seven shy of completing their tours), Thorsness and Johnson were shot down by a Mikoyan-Gurevich MiG-21 over North Vietnam while flying aircraft F-105F, AF Ser. No. 62-4447. He had flown the morning mission to the Hanoi area as Wild Weasel leader, then assigned himself as a spare aircraft for the afternoon mission because of a shortage of crews. One of Carbine flight aborted with radio problems, and Thorsness filled in as Carbine 03, leading the second element.

While still inbound over northwest North Vietnam, communications were disrupted when an ejection seat emergency beeper went off aboard one of the F-105s. Despite being observed by early warning radar locations, two MiG-21s approached Carbine flight from behind and unseen. Just as Thorsness got an instrument indication that the flight was being painted by airborne radar, he saw an F-105 going down in flames that eventually was identified as his own wingman, Carbine 04 (1st Lt Robert Abbott, in F-105D, AF Ser. No. 59-1726), shot down by an Atoll missile. Within a minute, his own aircraft was also hit with a heat-seeking missile fired by the MiGs.

Thorsness and Johnson ejected. Separated from each other by a ridge, they were the object of a three-hour rescue effort involving the entire strike force as a covering force. Two F-105D aircraft were directed by Crown to provide RESCAP (as Tomahawk flight) until the combat search and rescue (CSAR) forces could arrive on station. Both aircraft were hit by Atoll missiles from MiG 21s, with F-105D, AF Ser. No. 61-0130, piloted by Capt Joe Abbott being shot down, and wingman Maj. Al Lenski limping back to Thailand. In addition, one of the A-1 "Sandy" aircraft was hit while one of the rescue Jolly Greens developed hydraulic problems and had to abort, thus ending the CSAR mission. Poor communications, heavy MiG engagements and standard operating procedures which did not allow only one CSAR helicopter to remain on station, made the effort futile and all the men were captured. CSAR forces were again launched the next day but none of the downed airmen were located. The mission is described in great detail, including verbatim transcripts of radio transmissions, in both *Thud Ridge* and *Thud*, written by Col Jack Broughton, member of Waco flight and another of the RESCAP crews involved in the incident.

His uncooperativeness towards his captors earned him a year in solitary confinement and severe back injuries due to torture. The Medal of Honor was awarded to Thorsness during his captivity, but not announced until his release in 1973 to prevent the Vietnamese from using it against Thorsness, as was the Air Force Cross awarded to Capt Johnson for the same mission. Capt. Abbott was released from captivity on February 18, 1973, while Thorsness, Johnson, and 1st Lt Abbott were released on March 4, 1973, during Operation Homecoming.

Injuries incurred during the ejection and aggravated by the torture Thorsness was subjected to disqualified him medically from further flying in the Air Force and he retired on October 25, 1973 at the rank of colonel.

Col. Thorsness was a command pilot in the United States Air Force with 5,000 flying hours in L-21, T-6 Texan, T-28 Trojan, T-33 Shooting Star, F-84B and D Thunderjet, F-84F Thunderstreak,

F-100C and D Super Sabre, and F-105B, D, and F Thunderchief aircraft.

From 1979 to 1985, Thorsness served as Director of Civic Affairs for Litton Industries. He then served as a State Senator in Washington. Following his retirement, he served on the Board of Directors of the Congressional Medal of Honor Foundation. He moved from Catalina, Arizona, to Madison, Alabama, with his wife in early 2008 to be close to family.

In 2004, the University of Richmond announced the establishment of an endowed chair in leadership and ethics named in honor of Thorsness. The *Colonel Leo K. and Gaylee Thorsness Endowed Chair in Ethical Leadership* was funded by a $1,000,000 gift organized by W. Thomas Matthews, President and CEO of the Global Private Client Group at Smith Barney. Thorsness also served as Distinguished Leader in Residence at the Jepson School of Leadership Studies. The Thorsness chair is held by Donelson R. Forsyth, a social psychologist with expertise in group dynamics.

Thorsness died on May 2, 2017 in St. Augustine, Florida, at the age of eighty-five. According to his wife, Gaylee, the cause of death was leukemia. He is buried with full military honors at Arlington National Cemetery. In August 2020, Thorsness was named the Class Exemplar for the United States Air Force Academy's Class of 2023.

Thorsness was the Republican nominee against the incumbent Democrat George McGovern. It was believed that McGovern faced possible defeat for allegedly having neglected the state during his long 1972 presidential campaign. By May 1973, McGovern had already begun campaigning for reelection. Thorsness had just been repatriated after six years as a prisoner of war in North Vietnam. He accused McGovern of having given aid and comfort to the enemy and of having prolonged Thorsness' time as a POW. McGovern replied that if there had been no war, there would have been no POWs, and that everything he had done had been towards the goal of ending the war sooner. However, the war did not become a significant issue in the Senate race. Instead, the campaign was dominated by farm policy differences and economic concerns over the 1973–75 recession. Thorsness charged McGovern with being a "part-time senator" more concerned with national office and with spending over $2 million on his re-election bid, while McGovern labeled Thorsness a carpetbagger because he was originally from Minnesota. In a year in which Democrats were advantaged by the after-effects of the Watergate scandal, McGovern won re-election in November 1974 with 53 percent of the votes cast.

Thorsness settled in Seattle Washington, and was elected to the state senate on November 8, 1988. In accordance with Washington State law, he immediately became the senator from District 11 to serve the unexpired term of Avery Garrett, who died in April 1988. In January 1989, Thorsness took the oath of office for a four-year term. In the state senate Thorsness sponsored a bill dubbed the "Truth Bill" on March 3, 1990. The legislature unanimously passed the measure, SJM 8020, urging the Federal government to release information about 30,000 U.S. soldiers listed as either prisoners of war or missing in action in conflicts dating back to World War II. It further urged the United States Congress to pass a similar measure, HR3603, which would force the federal government to declassify information pertaining to over 30,000 missing American servicemen. In sponsoring the bill, Thorsness said that the government kept the information classified to protect intelligence sources, but that the sources are no longer useful because the conflict occurred decades ago. In 1992, he was an unsuccessful candidate in the Republican primary election for United States Senate. After a single term, Thorsness retired to Indianola, Washington.

WILLIAM WAYNE SEAY

POSTHUMOUS AWARD

Rank and Organization: Sergeant, 62nd Transportation Company, 7th Transportation Battalion, 48th Transportation Group; **Military Service Branch**: U. S. Army; **Conflict**: Vietnam War; **Medal of Honor Action Date**: August 25, 1968; **Medal of Honor Action Place**: Near Ap Nhi, Republic of Vietnam

Presentation: April 7, 1970 at The White House, presented by President Richard Nixon to his family

Accredited to: Montgomery, Montgomery County, Alabama

Born: October 24, 1948, Brewton, Escambia County, Alabama

Died: August 25, 1968, Republic of Vietnam

Buried: Weaver Cemetery (MH), Brewton, Alabama

Citation

For conspicuous gallantry and intrepidity in action at the risk of his life above and beyond the call of duty. Sgt. Seay distinguished himself while serving as a driver with the 62d Transportation Company, on a resupply mission. The convoy with which he was traveling, carrying critically needed ammunition and supplies from Long Binh to Tay Ninh, was ambushed by a reinforced battalion of the North Vietnamese Army. As the main elements of the convoy entered the ambush killing zone, they were struck by intense rocket, machine-gun, and automatic-weapons fire from the well-concealed and entrenched enemy force. When his convoy was forced to stop, Sgt. Seay immediately dismounted and took a defensive position behind the wheels of a vehicle loaded with high-explosive ammunition. As the violent North Vietnamese assault approached to within 10 meters of the road, Sgt. Seay opened fire, killing two of the enemy. He then spotted a sniper in a tree approximately 75 meters to his front and killed him. When an enemy grenade was thrown under an ammunition trailer near his position, without regard for his own safety he left his protective cover, exposing himself to intense enemy fire, picked up the grenade, and threw it back to the North Vietnamese position, killing four more of the enemy and saving the lives of the men around him. Another enemy grenade landed approximately three meters from Sgt. Seay's position. Again Sgt. Seay left his covered position and threw the armed grenade back upon the assaulting enemy. After returning to his position he was painfully wounded in the right wrist, however, Sgt. Seay continued to give encouragement and direction to his fellow soldiers. After moving to the relative cover of a shallow ditch, he detected three enemy soldiers who had penetrated the position and were preparing to fire on his comrades. Although weak from loss of blood and with his right hand immobilized, Sgt. Seay stood up and fired his rifle with his left hand, killing all three and saving the lives of the other men in his location. As a result of his heroic action, Sgt. Seay was mortally wounded by a sniper's bullet. Sgt. Seay, by his gallantry in action at the cost of his life, has reflected great credit upon himself, his unit, and the U.S. Army.

Epilogue

Seay joined the Army from Montgomery, Alabama in 1967, and by August 25, 1968, was serving as a Sergeant in the 62d Transportation Company (Medium Truck), 7th Transportation Battalion, 48th Transportation Group. On that day, a convoy of the 81 trucks escorted by eight Military Police gun jeeps left Long Binh Post for the 1st Brigade, 25th Infantry Division at Tây Ninh Combat Base. Traveling at the mandated convoy speed of 20 mph, the convoy passed through the village of Ap Nhi. Approximately 11:45, a battalion sized enemy force opened fire from the Ben Chu Rubber Plantation on their right. The lead 30 5-ton cargo trucks escaped the kill zone, but the enemy set two fuel tankers on fire at the front of the convoy and two ammunition trailers on fire at the rear trapping the convoy in between. The enemy fire forced the crews of the trucks and gun jeeps to quickly dismount and take cover leaving their M60 machine guns on their jeeps. The enemy then made a concerted effort to destroy the radios and M-60s on the gun jeeps, but one MP managed to get to his jeep and put in a call for help about 15 minutes into the fight. Meanwhile, SGT Seay and SP4 David Sellman took cover behind the trailers of their rigs and fought back. Seay left the cover of his position several times to toss hand grenades back and was finally wounded in the wrist. The nearest mechanized infantry security forces approached with M113 armored personnel carriers but the intense enemy fire would not allow them to enter the kill zone, so they established casualty collection points at both ends of the kill zone. Truck drivers and MPs either fought in small pockets like Seay and Sellman or maneuvered to the safety of the ends of the convoy. Seay found LT Howard Brockbank and four other soldiers where they treated his wound. He continued to fight back enemy assaults until he died from loss of blood. The ambush lasted for over nine hours and for his actions during the battle, Seay was awarded the Medal of Honor.

Seay, aged 19 at his death, was buried in Weaver Cemetery, Brewton, Alabama.

The Seay Army Reserve Center on Commanders Drive in Mobile, Alabama was dedicated and named in honor of William Seay for his heroic service. Lt. Col. (dec) Arthur Jones Jr. was responsible for this name assignment and dedication. The Seay Auditorium at U.S. Transportation Command, Scott AFB, Illinois, is also named in his honor.

BENNIE G. ADKINS

Rank and Organization: Sergeant First Class, Detachment A-102, 5th Special Forces Group, 1st Special Forces; **Military Service Branch**: U. S. Army; **Conflict**: Vietnam War; **Medal of Honor Action Date**: March 9-12, 1966; **Medal of Honor Action Place**: Camp A Shau, Republic of South Vietnam

Presentation: September 15, 2014 at The White House, presented by President Barack Obama

Accredited to: Waurika, Jefferson County, Alabama

Born: February 1, 1934, Waurika, Jefferson County, Alabama

Died: April 17, 2020

Buried: Arlington National Cemetery, Virginia

Citation

Sergeant First Class Bennie G. Adkins distinguished himself by acts of gallantry and intrepidity at the risk of his life above and beyond the call of duty while serving as an Intelligence Sergeant with Detachment A-102, 5th Special Forces Group, 1st Special Forces, during combat operations against an armed enemy at Camp A Shau, Republic of Vietnam, from March 9 to 12, 1966. When the camp was attacked by a large North Vietnamese and Viet Cong force in the early morning hours, Sergeant First Class Adkins rushed through intense enemy fire and manned a mortar position continually adjusting fire for the camp, despite incurring wounds as the mortar pit received several direct hits from enemy mortars. Upon learning that several soldiers were wounded near the center of camp, he temporarily turned the mortar over to another soldier, ran through exploding mortar rounds, and dragged several comrades to safety. As the hostile fire subsided, Sergeant First Class Adkins exposed himself to sporadic sniper fire while carrying his wounded comrades to the camp dispensary. When Sergeant First Class Adkins and his group of defenders came under heavy small arms fire from members of the Civilian Irregular Defense Group that had defected to fight with the North Vietnamese, he maneuvered outside the camp to evacuate a seriously wounded American and draw fire, all the while successfully covering the rescue. When a resupply air drop landed outside of the camp perimeter, Sergeant First Class Adkins, again, moved outside of the camp walls to retrieve the much-needed supplies. During the early morning hours of March 10, 1966, enemy forces launched their main attack and within two hours, Sergeant First Class Adkins was the only man firing a mortar weapon. When all mortar rounds were expended, Sergeant First Class Adkins began placing effective recoilless rifle fire upon enemy positions. Despite receiving additional wounds from enemy rounds exploding on his position, Sergeant First Class Adkins fought off intense waves of attacking Viet Cong. Sergeant First Class

Adkins eliminated numerous insurgents with small arms fire after withdrawing to a communications bunker with several soldiers. Running extremely low on ammunition, he returned to the mortar pit, gathered vital ammunition and ran through intense fire back to the bunker. After being ordered to evacuate the camp, Sergeant First Class Adkins and a small group of soldiers destroyed all signal equipment and classified documents, dug their way out of the rear of the bunker and fought their way out of the camp. While carrying a wounded soldier to the extraction point he learned that the last helicopter had already departed. Sergeant First Class Adkins led the group while evading the enemy until they were rescued by helicopter on March 12, 1966. During the 38-hour battle and 48 hours of escape and evasion, fighting with mortars, machine guns, recoilless rifles, small arms, and hand grenades, it was estimated that Sergeant First Class Adkins had killed between 135 and 175 of the enemy while sustaining 18 different wounds to his body. Sergeant First Class Adkins' extraordinary heroism and selflessness above and beyond the call of duty are in keeping with the highest traditions of the military service and reflect great credit upon himself, Detachment A-102, 5th Special Forces Group, 1st Special Forces and the United States Army.

Epilogue

Adkins was born in Waurika, Oklahoma and was drafted in 1956. He was assigned to a garrison unit in Germany, with a follow-on assignment to the 2nd Infantry Division, Fort Benning, Georgia. After attending Airborne School, he volunteered for Special Forces in 1961, serving with Special Forces for more than 13 years with the 7th, 3rd, 6th and 5th Special Forces Groups (Airborne). During that time he deployed to the Republic of Vietnam three times between 1963 and 1971. In April 1967, Adkins was awarded the Distinguished Service Cross for his actions with Detachment A-102 during his second tour in Vietnam. After Vietnam, Adkins was assigned to Fort Huachuca. Graduating in the third class of the Sergeant Major Academy, he returned to the Special Forces at Fort Bragg, then went to Fort Sherman and led training at its Jungle Operations Training Center. Adkins finally retired from the Army in 1978.

After the Army, Adkins earned a bachelor's and two Master's degrees from Troy State University. He operated his own accounting company, and taught classes at Southern Union Junior College and

Auburn University. On May 12, 2017, Troy University Chancellor Jack Hawkins, Jr. awarded Adkins an honorary doctorate of laws.

In March 2020, Adkins was hospitalized with COVID-19. He was admitted to the intensive care unit and put on a ventilator after experiencing respiratory failure. He died on April 17, 2020, from complications of the virus at the age of 86.

From 2002, the U.S. Army reviewed all 6,500 recipients of the Distinguished Service Cross to see if any recipients had actually performed actions worthy of the Medal of Honor; this led to two dozen medal upgrades in March 2014. In 2013, as part of the National Defense Authorization Act for Fiscal Year 2014, the Senate Armed Service Committee passed a provision removing the time limit for Donald P. Sloat and Adkins. On September 15, 2014, President Obama awarded the Medal of Honor to Adkins as an upgrade of his 1967 Distinguished Service Cross. During that ceremony, the Medal of Honor was awarded posthumously to Sloat and American Civil War army officer Alonzo Cushing. Adkins was also inducted into the Pentagon's Hall of Heroes.

SELECT BIBLIOGRAPHY/RECOMMENDED READING

Beyer, W. F. & O. F. Keydel. Deeds of Valor. New York: Longmeadow Books, 1994.

Gumpertz, Sydney G. *The Jewish Legion of Valor: The Story of Jewish Heroes in the Wars of the Republic*. Self-published, 1934.

Hardy, Gordon (ed.) *Above and Beyond: A History of the Medal of Honor from the Civil war to Vietnam*. Boston: Boston Publishing Co., 1985

Lemon, Peter C. *Beyond the Medal: A Journey from Their Hearts to Yours*. Golden, CO: Fulcrum, 1997.

McChristian, Douglas. *The U. S. Army in the West, 1870 – 1880*. Norman: University of Oklahoma Press, 1995.

O'Neil, Bill. *Fighting Men of the Indian Wars*. Western Publications, 1992.

Proft, R. J. (ed.) *United States of America's Congressional Medal of Honor Recipients*. Columbia Heights, MN: Highland House, 1998.

Ward, Geoffrey C. *The Civil War. New York*: Alfred A. Knopf, 1990.

Lee, Irvin H. *Negro Medal of Honor Men* New York: Dodd, Mead, 1967

Smith, Larry. Beyond Glory: *Extraordinary Stories of Courage from World War 2 to Vietnam War*.

Wallace, Mike. Medal of Honor: *Profiles of American Military Heroes from the Civil War to the Present*. New York: Artisan Press.

Jordan, Kenneth N. *Yesterday's Heroes: 433 Men of World War II Awarded the Medal of Honor 1941-1945*.

Del Calzo, Nick & Collier, Peter. Medal of Honor: portraits of Valor Beyond the call of Duty. New York: Artisan.

Wesley, Charles H & Patricia W. Romero. *Afro-Americans in the Civil War, from Slavery to Citizenship*. Pennsylvania: The Publishers Agency, Inc., 1978.